T0323943

"Joyce Shaffer's chapter on Neurobics reveals her mastery of the subject of neuroplasticity. It covers the literature on the impact of music and dance on the agility of the aging or injured brain, and remains eminently readable, for scientist and layperson alike."

Anne Fanning, MD, *Professor Emeritus, Faculty of Medicine, University of Alberta, WHO Medical Officer for Global TB Education from 1998 to 1999*

"Dr. Shaffer's wise and insightful book illuminates her depth of expertise and passion for sharing sound and proven guidance for those of us to wish to live a more productive, fulfilling, and enjoyable life with a song in our hearts throughout our senior years."

John Kander II, *Executive Director, Music Mends Minds,* **310.567-8464/ john@musicmendsminds.org**

"This book will change lives of millions of people, giving them awareness particularly in the third world countries where there is less research and knowledge. Works like this encourage people to transform their lives for prevention, delay onset and reverse dementia for a healthy brain."

Dr. Irfan Gul Magsi, *MBBS (AKUH), Member of Parliament, Minister for Revenue and Religion Affairs Government of Pakistan*

"Dr. Joyce offers a comprehensive and groundbreaking exploration of our brain's incredible plasticity and the science behind positive aging. It's a must-read for anyone interested in the latest advancements in neuroscience and how to apply them for a healthier, more vibrant life."

Kelley Keehn, *Best-selling author of 11 books, including Talk Money to Me*

"I have had the pleasure of knowing and working with Dr. Joyce Shaffer, PhD, for the past 8 years through Music Mends Minds' Global Initiative Platform. As we partnered to spread the joy of music to our aging population affected by neurodegenerative disorders, it has become evident that Dr. Shaffer's lifetime of scientific expertise highlights her as a trailblazer in substantiating the role of arts on brain health. Our collaboration of science and music has uplifted the quality of lives for our seniors, and this book is a gift with information to help you elevate your brain function."

Carol Rosenstein, *President and Founder of Music Mends Minds, Proud Member of Westwood Village Rotary Club*

"I enjoyed your thoughts and research on the psychological effects of Music & Dance on the brain. As a former Parks and Recreational Professional, I saw how Music and Dance programs had a direct effect on the brain and memory loss!

As the Founder of a Rotary Action Group called ADRAG (Alzheimer's/ Dementia Rotary Action Group), I saw those with a form of Dementia enjoying Music & Dance and reacting – it was like a form of medicine.

Dr. Joyce Shaffer is a writer who cares about and shares with others. She truly exemplifies the spirit of serving others by writing books of interest."

David I. Clifton, *Founder, Alzheimer's/Dementia Rotary Action Group*

"I have always been in awe of Dr. Joyce's contagiously affectionate personality. She is my mentor, a great friend, and an empathetic listener.

I feel honored to be able to read this amazing piece of work which will stand as landmark in improving brain health, concept of neuroplasticity, and healthy brain aging.

This book is based on extensive research and enlightened workmanship. The chapters where Dr. Joyce writes about Positive Agers, Super Agers, and Centenarians Superior Successes are truly inspiring.

The chapter about Love, Touch, Talk, and Listen felt truly relatable and comforting to me as a mother.

All in all, this book is the need of time. I truly hope this amazing piece of research reaches the masses. It's a breakthrough towards future of Brain health, to prevent, delay onset, and even reverse symptoms of Dementia."

Sana Irfan Magsi, *Masters in International Relations, Masters in Organizational Leadership, Business Development Manager, Sales Team Lead*

Hacking Neuroplasticity

Neuroplasticity is the ability of neural networks in the brain to change through growth and reorganization. It is when the brain is rewired to function in some way that differs from how it previously functioned.

How does aging affect neuroplasticity? As we grow older, plasticity decreases to stabilize what we have already learned. What influence does the aging process have on memory? Forgetfulness can be a normal part of aging. As people get older, changes occur in all parts of the body, including the brain. As a result, some people may notice that it takes longer to learn new things, they don't remember information as well as they did, or they lose things like their glasses. This book on evolving neuroscience is unique in its lifespan focus on driving neuroplasticity in a positive direction to influence the Flynn effect of increasing human intelligence as the preferred way to prevent, delay onset, and/or reverse dementia. It considers potential impact from the first moments of life through end of life. It includes intergenerational activities. Its inclusion of centenarians and supercentenarians provides examples of "Super Agers" who have maintained and/or increased neurocognitive capacity, often with a health span that approximated their vigorous longevity. It discusses the use of the artificial intelligence (AI) revolution to refine, personalize, and broaden our global reach to enhance the Flynn effect as the preferred effort to improve global statistics on neurocognitive functioning at any age. Driving neuroplasticity in a positive direction at all ages is urgent.

With this book's focus on evidence-based interventions at any age which can have physical, emotional, neurobiological, neurochemical, immunological, and social health benefits, it is a unique overview and application of evolving neuroscience to address the UN/WHO Decade of Action for Healthy Ageing for All.

Hacking Neuroplasticity
How AI Can Help
Your Healthy
Aging Brain

Dr. Joyce Shaffer

Foreword by Tom Lawry, author of
AI in Health and Hacking Healthcare

Routledge
Taylor & Francis Group

NEW YORK AND LONDON

First published 2025
by Routledge
605 Third Avenue, New York, NY 10158

and by Routledge
4 Park Square, Milton Park, Abingdon, Oxon, OX14 4RN

Routledge is an imprint of the Taylor & Francis Group, an informa business

© 2025 Joyce Shaffer

The right of Joyce Shaffer to be identified as author of this work has been asserted in accordance with sections 77 and 78 of the Copyright, Designs and Patents Act 1988.

All rights reserved. No part of this book may be reprinted or reproduced or utilised in any form or by any electronic, mechanical, or other means, now known or hereafter invented, including photocopying and recording, or in any information storage or retrieval system, without permission in writing from the publishers.

Trademark notice: Product or corporate names may be trademarks or registered trademarks, and are used only for identification and explanation without intent to infringe.

ISBN: 9781032611723 (hbk)
ISBN: 9781032611716 (pbk)
ISBN: 9781003462354 (ebk)

DOI: 10.4324/9781003462354

Typeset in Adobe Garamond Pro
by codeMantra

Contents

Preface

Neuroplasticity is one of the greatest gifts of our existence. Our brains are plastic! Our brains are an artwork in progress!

The rapidly changing database of evidence-based interventions to help us be the best possible artists in the evolving artwork of our neuroplasticity enhancements is a gift. It is empowering because building bigger, better brains has been documented in evolving neuroscience described herein. Artificial intelligence can add superpowers to our joint efforts in the Intelligent Neuroplasticity Revolution.

Using these evidence-based interventions could be good for business, education, healthcare, government, philanthropy, and personal development as described herein. *Not* using these evidence-based interventions leaves all of us at risk for detrimental outcomes also described herein which include escalating healthcare costs.

These changes are urgently needed and can be facilitated in ways that people enjoy. Enjoying it is essential because people want to do what is fun, easy, portable, culturally adaptable, and free or low cost.

Additional good news is that many of these evidence-based interventions also improve physical, mental, neurobiological, emotional, career, and prosocial health and well-being. That could influence significant savings in healthcare costs. It could also help promote improved quality of life with more civil discourse and peaceful coexistence.

Positive Psychology[2] is the focus of this writing because negative emotions can be detrimental to neurons, whereas positive emotions can increase the health of brain cells. Also, people are more likely to do what is fun. It is fortunate that several of these evidence-based interventions are fun, easy, portable, culturally adaptable, and free or low cost. That is pertinent across the lifespan.

My studies across decades on neuroplasticity began with the focus on increasing the Flynn[3] effect of rising IQ scores found in recent decades in

34 countries. At first, my studies were pursued for the joy of learning and helping healthy people. My mission was to empower the gifted who have a track record of service for good. It was learning of the frequency that a new person among our global citizens was diagnosed with dementia that gave me pause for thought.

The Intelligent Neuroplasticity Revolution is urgent. Hacking Neuroplasticity for Your Healthy Aging Brain is urgent. Using our best human efforts with artificial intelligence to hasten an international effort to enhance neuroplasticity, increase prosociality, and increase human intelligence as a preferred way to prevent, delay onset, and/or reverse dementia is urgent. Evolving neuroscience is increasingly encouraging and empowering. Together, we can be "superpowers for change" in our efforts to make the World a better place for ALL.

Acknowledgments

Thank you. Terima Kasih. Gracias. Grazie. Merci. Xie xie. M-Goi. & more!

Teamwork makes our Intelligent Neuroplasticity Revolution a global effort of growing international reach in the interest of building bigger, better brains for ALL. Although my name is on the cover, if I included everyone to whom I am indebted, this book might be too heavy to carry. In addition to the individuals mentioned below, there are many other gracious people to whom I am indebted for their influence and input on this book; and most of you know who you are!

Marian Diamond, PhD, was Professor Emeritus at the University of California, Berkeley. Her research provided the first evidence of brain changes with experience, first evidence of brain changes with enrichment, and a list of influences that drive neuroplasticity in a positive direction. Her proof of neuroplasticity preceded the use of that word. Thus, she is the Mother of Neuroplasticity. Arguably, she is also the Mother of Positive Psychology based on her research you can read herein. She asserted that these neuroplastic changes were possible in humans through the age of 100. She described related evidence that she found in the brain of Albert Einstein.[4]

Marian Diamond sat in the chair beside me when she was a major presenter for our Washington State Psychological Association. The information she taught us changed my life forever. From that moment forward, I will remain indebted to her for her erudite influence. Dr. Diamond's research value cannot be overstated. Please join me in sharing her research and perspectives everywhere possible in the interest of making our World a better place for ALL.

Fred H. Gage, PhD, Adler Professor at The Salk Institute, provided another very empowering paradigm shift by proving neurogenesis. He added to our "superpowers for change" by providing evidence that the birthrate of new neurons and the percentage of survival of these new brain cells can be increased by lifestyle choices.

Tom Lawry is the best-selling author of *Healthcare – How AI and the Intelligent Health Revolution Will Reboot an Ailing System.*[1] His book is so superior that I keep rereading it and learning something new each reread!

He served as a former National Director for AI for Health and Life Sciences, Director of Worldwide Health, and Director of Organizational Performance for Microsoft. He is a frequent keynote speaker on AI.

Tom Lawry cannot be thanked enough for his essential education of others on this topic AND for his perspective on empowering others to harness "the power of AI and the Intelligent health Revolution to create a sustainable system that focuses on keeping all citizens healthy while caring for them when they are not." His perspective and ongoing efforts help us make the World a better place for ALL.

Rudolph E. Tanzi, PhD, Joseph P. and Rose F. Kennedy Professor of Neurology, Harvard Medical School Vice-Chair, Neurology Department, Massachusetts General Hospital Director, Genetics and Aging Research Unit, Massachusetts General Hospital.

Rudolph Tanzi's lab has been involved in identifying three Alzheimer's genes. His TEDx video[5] is worth watching more than once for his science and perspective. It ends with Dr. Tanzi playing keyboards while Chris Mann sings the Alzheimer's song, *Remember Me,* which was composed by Dr. Tanzi. The power of *Remember Me* as performed at the end of this video was so profoundly touching that I've watched it many times. Dr. Tanzi's neuroscience contributions coupled with his brilliance in composition and playing piano music make him a very multifaceted gift to our World.

Dr. Tanzi's collaboration and work with MusicMendsMinds.org is much appreciated. We agree with him that music is medicine.

Liliana Alberto, MD, estudiad Médico cirujano at Universidad Nacional de Córdoba. Retired Pediatric Pulmonologist, and District Governor of Rotary International District 4945.

Just as my focus on enhancing neuroplasticity at any age in *healthy* people made a major shift in response to the global statistics on dementia, Dra. Alberto was a guest in my Bellevue Rotary Club, District 5350. We cofounded an international effort to prevent, delay onset, and/or reverse dementia. She teaches from everything I write with her activities in Argentina, Uruguay, Paraguay, Chile, Brazil, and beyond. She arranged for me to join her efforts for two weeks in Argentina and arranged events presenting to about 2000 people.

She has also been instrumental in combining our efforts with the works of MusicMendsMinds.org. For World Brain Day 22 July 2023, we had activities on 4 continents! Liliana provides serious winds beneath my wings. When I suggested that I stop what I'm doing and get as much activity in North America as she has accomplished in South America, she adamantly said: "No, keep doing what you are doing." So, I enjoy telling friends that Liliana told me to stay in the library where I belong.

Carol Rosenstein enjoyed a career as a chiropractor, earned a master's in clinical psychology, and practiced mind/body medicine for many years. She is a proud member of Westwood Village Rotary Club, District 5280 AND the President and Founder of the Music Mends Minds organization for restoring the rhythm of life. Their activities in many communities in several countries have a global reach using music for healing body and mind.

Their website MusicMendsMinds.org provides many beneficial activities, including podcasts and virtual events for individuals who cannot travel somewhere to join a drum circle; and their events are intergenerational. They were instrumental in getting activities on four continents for World Brain Day 22 July 2023!

John Kander II is Executive Director of Music Mends Minds, a ten-year-old Los Angeles-based, global non-profit that creates free music groups for seniors, many of whom have neurodegenerative challenges. He feels fortunate that his award-winning media work often influenced the national conversation while his volunteer work has aided several common good organizations. He is proud to be bringing the joys and medicine of music to those who need it most.

Eric J Topol, MD, cardiologist and author, is Director and Founder of Scripps Research Translational Institute, as obtained from Eric Topol – Wikipedia.

Although I have never met Dr. Topol, I want to express gratitude for how much I've learned from his TED talks. I have shared them as they fuel my passion for Hacking Neuroplasticity and an Intelligent Neuroplasticity Revolution.

David Yu, MD, provided the many delightfully expressive illustrations in this book which were produced while he was a medical student at the University of Washington in Seattle. Brilliant clinician and artist!

CWO Charles T. Shaffer during his tour of duty with the US Navy traveled home to ask how my applications for nurses' training were going. When I said I was not applying because I didn't have the money, in an instant my brother Charles said: "You tell me how much the 3 years will cost, and I will send you the money."

No mention of a loan, of repayment, nor of interest charged on those funds. Without my brother Charles funding that schooling, it would never have been possible to continue my education through a doctorate in psychology, two master's degrees in psychology, and other credentials; it would never have been possible to read, understand, publish, and teach neuroplasticity. That is why I give total credit to my brother Charles for my entire education and publications.

An individual who must remain unidentified gets similar credit for my growing passion for and knowledge about neuroplasticity. After 15 years in Special Education in public schools with repetitive math and reading remediation classes, those skill deficits persisted. A short period in a community college with similar remediation efforts was equally ineffective.

Then, a computerized math training program became available. As described herein, that improved this individual's math skills to the extent that their *speed of processing accurate mental math responses very significantly exceeded* those of specialists with advanced degrees. Observing that set the stage for my serious appreciation of the research described herein on the very substantial benefits of computerized cognitive training for neurocognitive and other functional gains as shared herein.

Kristine Rynne Mendansky is Senior Editor at Taylor and Francis. She has been graciously helpful in answering many questions and patient with my progress. Am *very* grateful to Tom Lawry for introducing us. Am *very grateful* to Kris for working with me.

Jaime Dubè very graciously provided her excellent illustration of the Rescuing Hug. It says more than 10,000 words!

Gratitude comes easily to many more people who have made this book possible, saying thank you is heartfelt. The Intelligent Neuroplasticity Revolution is urgent and can be fun, easy, portable, free or low cost, and culturally adapted. Here's wishing you many decades more in your joy of Hacking Neuroplasticity for your Intelligent Neuroplasticity Revolution to help Your Healthy Aging Brain.

The University of Washington and Harborview Medical Center

It has been a privilege as well as a very sobering experience to have worked with individuals in Harborview. Our vulnerable people who are so seriously struggling have increased my passion for reading, writing, teaching, and sharing evolving neuroscience. Working for the University of Washington increases my access to this important database. Hacking neuroplasticity for the benefit of all is one of the ways that we can make our world a better place.

Lyalla Magsi

Honor student, artist, poet. Won many international medals in global debate competitions.

As a creative artist, Lyalla help refine and upgrade some of my graphics.

References

1. Lawry T. (2023). *Hacking Health Care: How AI and the Intelligence Revolution Will Reboot an Ailing System*. New York: Routledge.
2. Peterson C. (2008) What Is Positive Psychology, and What Is It Not? *Psychol Today*, posted online May 16, 2008; https://www.psychologytoday.com/us/blog/the-good-life/200805/what-is-positive-psychology-and-what-is-it-not? (Accessed online July 2, 2023).
3. Flynn JR. (2018). Reflections about Intelligence over 40 Years. *Intelligence*, 70: 73–83. https://doi.org/10.1016/j.intell.2018.06.007
4. Diamond MC, Scheibel AB, Murphy Jr GM & Harvey T. (1985) On the Brain of a Scientist: Albert Einstein. *Exp Neurol*, Apr; 88(1): 198–204. https://doi.org/10.1016/0014-4886(85)90123-2.
5. Curing Alzheimer's with Science and Song | Rudy Tanzi & Chris Mann | TEDxNatick (youtube.com)

About the Author

Joyce Shaffer serves as an expert on psychiatric and medical matters for court systems. She earned her doctorate in psychology and second master's degree in psychology from Hofstra University in New York; a bachelor's degree and first master's degree in psychology from Towson State University in Baltimore, Maryland; and completed nurses' training at the Thomas Jefferson School of Nursing in Philadelphia, Pennsylvania.

For the University of Washington in Seattle at Harborview Medical Center as a court evaluator, she works with clinicians, patients, their families, and caregivers to serve in ways that put patients first. She has worked on inpatient units in Alaska, Pennsylvania, Maryland, New York and Washington states in the USA. Her clinical evaluations and interventions in hospitals and when seeing people privately have included providing information on enhancing neuroplasticity in ways that can influence promoting academic, business, physical, mental, emotional, career, and social health.

Her studies of driving neuroplasticity in a positive direction began with a focus on increasing the Flynn effect of rising IQ scores found in recent decades in 34 countries. Her goal has always been to bring those same benefits of enhancing neuroplasticity to more global citizens. That evolved to include being cofounder of an international effort to prevent, delay onset, and/or reverse dementia. Evolving neuroscience is increasingly encouraging and empowering. She invites participation in celebrating World Brain Day on 22 July every year everywhere.

She has published numerous articles and is a frequent conference speaker on evolving neuroscience evidence-based lifestyle choices that can improve brain chemistry, architecture, and performance from in utero and throughout a vigorous longevity. She is grateful to have the opportunity to add working with artificial intelligence to expedite, refine, and personalize ongoing efforts to drive neuroplasticity in a positive direction for the HEALTH of it in our Intelligent Neuroplasticity Revolution.

Foreword

In the ever-evolving landscape of neuroscience, few concepts have captivated the scientific community and the general public alike as much as brain plasticity. The brain's remarkable ability to rewire itself, to form new neural connections, and to adapt in response to experience is nothing short of miraculous.

This new book by Joyce Shaffer, PhD, ABPP, provides a lot to learn and consider about the human brain.

But if you are in a hurry, let me net out what you should know: The human brain is an awesome organ. Weighing an average of 3 pounds, it has 100 billion neurons and 100,000 miles of capillaries.[1] It's what lifted homo-sapiens to the top of the food chain.

You're probably not conscious of how your brain is working right now:

> *As soon as you saw this page, something started happening inside your cranium. Electrical transmissions from your eyes triggered neuro-responders in your brain. Traveling at speeds up to 268 miles per hour, these signals stimulated your brain to begin recognizing patterns known to you as letters, words, and punctuation.*
>
> *Based on other cognitive capabilities, these words encapsulate and trigger retained knowledge from previous patterning that is unlocked based on the unique sequence of patterns on the pages of this book.*

Despite accounting for only about 2% of the body's weight, the brain consumes approximately 20% of the body's energy. This high energy consumption is necessary to maintain the brain's functions, including its extensive communication network and the constant activity of neurons.

The human brain is so smart that it has invented ways to outsource certain human intelligence capabilities to machines. Today, we call this artificial intelligence (AI) which is the field that I work in.

Not only was AI developed by humans, but it's also delivered through human-made elements such as silicon, plastics, and code.

Natural human intelligence (NHI) however is very different. It's delivered through a blob of protoplasm that AI pioneer Alan Turing once described as an organ "having the consistency of cold porridge."

AI has recently become the "shiny object" in the fields of science and technology. And while we continue to see AI evolve, its capabilities remain limited compared to the range of things the human brain can do.

One of the most remarkable and most misunderstood features of the brain is its plasticity, or the ability to reorganize itself by forming new neural connections throughout life. This adaptability is crucial for learning, memory, and recovery from brain injuries. For example, when one part of the brain is damaged, other parts can sometimes compensate.

Brain plasticity plays an important role in our health, happiness, and success at all stages of life. It's an especially important area to understand and leverage when considering the health and well-being of older adults.

Today, in America, birth rates continue to decline. Meanwhile, 10,000+ baby boomers turn 70 every day. And despite things like the COVID pandemic, there is growing evidence that we are at the beginning of a longevity revolution. In the not too distant future, many citizens can be expected to live well into their 100s.

In case you think this sounds far-fetched, consider the fact that lifespan has more than doubled in the last 100 years.[2]

In many ways, we have reached a point where longevity has become humanity's new frontier.

As we extend the amount of time the average human lives, the real issue is not lifespan but healthspan. Lifespan is the total number of years we live. Healthspan is how many of those years we have an acceptable level of health and well-being.

Rather than worrying about dying too young, many people today worry about living too long without sufficient quality of life. In the USA, the difference between life expectancy, or lifespan, and healthy life expectancy, or healthspan, is almost ten years.[3]

While there are many factors to consider in closing this gap, brain health or neuroplasticity is one of the top areas for consideration and further study.

While numerous studies are cited in this book, a randomized controlled clinical trial was recently published that demonstrates for the first time how an intensive lifestyle intervention, without drugs, may significantly improve cognition and function in many patients with mild cognitive impairment or early dementia due to Alzheimer's disease.[4]

Increasing evidence like this links lifestyle factors with the onset and progression of dementia, including Alzheimer's disease. These include unhealthful diets, being sedentary, emotional stress, and social isolation. The results of this study suggest that brain plasticity is a critical factor in relation to how changing life's daily habits has the potential to affect brain health.

While there is much more for us to understand, there is a growing body of evidence that challenges the long-held notion that the adult brain is immutable. Instead, we are seeing studies and data that offer a more hopeful perspective: that our brains are dynamic, capable of growth, and change at any age.

Brain plasticity is a remarkable feature of the human brain that underpins our ability to learn, adapt, and recover. Its importance to health and well-being cannot be overstated, as it influences cognitive development, recovery from injuries, mental health, aging, and adaptability. By understanding and harnessing the power of neuroplasticity through lifestyle choices and therapeutic interventions, we have the potential to broaden our understanding and learnings to enhance their brain health, resilience, and the overall quality of life for all.

Tom Lawry
Author of *Hacking Healthcare*

Notes

1 Emma Scott, How much does the human brain weigh? Med Health Daily, https://www.medhealthdaily.com/how-much-does-the-human-brain-weigh/
2 Max Roser, Twice as long – life expectancy around the world, Our World in Data, October 8, 2018. https://ourworldindata.org/life-expectancy-globally
3 Ken Dychtwald, Lifespan versus Healthspan, September 2, 2021, LinkedIn. https://www.linkedin.com/pulse/lifespan-versus-healthspan-ken-dychtwald/
4 Ornish, D., Madison, C., Kivipelto, M. *et al.* Effects of intensive lifestyle changes on the progression of mild cognitive impairment or early dementia due to Alzheimer's disease: a randomized, controlled clinical trial. *Alz Res Therapy* **16**, 122 (2024). https://doi.org/10.1186/s13195-024-01482-z

1

Our Brains Are Plastic … Our Brain Artworks in Process

Our brains are plastic. The media has been overloaded with all the negative and sobering evidence about how humans cause brain damage. Drinking too much alcohol. Coping poorly with stress. Too little exercise. Improper and inadequate sleep. Traumatic brain injury. The list goes on.

Fortunately, neuroscientists have begun to bring us more positive, more encouraging, *and* more empowering news. There's a growing tsunami of research providing evidence-based interventions that are associated with *enhancing* our brains. It's captured most succinctly in the title of the book *Enriching Heredity*[1] written by Marian Diamond, PhD, who was a pioneer in proving how many ways we can have that very positive impact on our brains. That surely makes her the Mother of Neuroplasticity, the first to show that our brains are plastic.

Since hearing Dr. Diamond describe how much we can influence *Enriching Heredity*, her research goes everywhere I go in writing, speaking, and designing my day. Each of us is the artist with the most impact on changing the evolving artwork of our plastic brains as we influence how our brains change across our lifespan. That is empowering.

I say "the most impact" because there is some truth to the saying that we become who we hang out with. Emotions are contagious. People who know things we do not know can help us learn. Each of us has our own version of how that plays out in our lives.

In my youth, I did not quite fit that mold of becoming who I hung out with. As number seven of seven children born on a farm, six of whom ran away from home, several of whom did not complete high school, my

DOI: 10.4324/9781003462354-1

attending university was aberrant. A major criticism shouted at me in my youth was: "All you *EVER* want to do is *READ!*"

That was years before I learned that all I really want to do is *read, write, and teach* neuroplasticity because we have so many ways to benefit from knowing that our brains are plastic and that we are the artisans influencing these brain artworks in process. Our life journey can include *Enriching Heredity.*

Flynn Findings for Future FUN

The Flynn effect suggests that we must be doing something right since increases in human intelligence have been documented in recent decades in 34 countries.[2] Even Flynn had his critics, perhaps most notably Arthur Jensen who objected to some aspects of the works of Flynn and his colleagues. Undeterred, Flynn modeled a wise perspective.[3]

> I will detail what I learned from my long debate with Arthur Jensen. He was irreplaceable in the sense that you often learn more from a thinker who challenges opinions you tend to take for granted. They shake you out of your dogmatic slumbers – force you to defend assumptions made unreflectively that cry out for clarification.
>
> *(Flynn, 2018)[3]*

The finding of Flynn and colleagues that human intelligence has been increasing in recent decades is known as the Flynn effect. It empowers us. What sort of future can we envision for self and others? Can our focused efforts increase the Flynn effect? If so, by how much? For whom? How rapidly? Which neurocognitive skills can be expected to improve? What evidence-based interventions might have the best, strongest, broadest effects? Where shall we begin?

Those questions are where the fun begins. The extensive overview of research written herein provides us with so many evidence-based interventions associated with neurocognitive gains that we can laugh our way through using the entire multiple-choice list of interventions that we have personalized using artificial intelligence and reviewed with our healthcare provider.

Laughing our way through is spot on as you will read later. But first, a bit more about the Flynn effect.

A recent review of datasets from 72 countries, including 299,155 participants, supported the Flynn effect of rising IQ scores.[4] Younger generations and middle-income countries had stronger Flynn effects.

Our Smorgasbord to Celebrate

Being alive in the age of technology gives us real-time access to evolving neuroscience. That continues to be a huge gift in making it so much more efficient for me to pursue my passion for reading, writing, and teaching neuroplasticity. Another cause for celebration is that stellar neuroscientists continue to refine, modify, and add to our many options of theme and variation on evidence-based interventions associated with driving neuroplasticity in a positive direction at any age and regardless of levels of ability. Many of those are discussed in later chapters in this book.

Positive emotions are high on that list of evidence-based options. They increase the neurochemicals that enhance brain health and decrease the stress chemicals which can, if prolonged, have a negative impact on neuroplasticity. Since emotions are contagious, these brain enhancing benefits of neurochemicals that are associated with your positive emotions could be similarly beneficial for individuals who can witness and share your happiness and laughter. In serving self by being cheerful, you share the healthy brain aging benefits with people who spend time being happy with you.

Gelotology is the study of laughter with recent research suggesting prescribing laughter to improve healthy aging.[5] Laughter is low risk, available to everyone, easily elicited, acceptable in many settings especially if in celebration of a person in some way, has few if any side effects, and might improve general health and longevity.

Artificial intelligence used with human creativity and wisdom is another huge reason for celebration. Human brains created artificial intelligence and human brains can enhance neuroplasticity by using it wisely as will be enlarged upon herein.

More Elders, Less Dementia Incidence

The Nun Study[6] combined annual evaluations of cognitive and physical function with postmortem neuropathologic evaluations. The "gold standard" for healthy aging, Sister Matthia, continued to be "happy, productive, and vivacious" to the end of her life just weeks short of 105, having "enjoyed more than 100 years of dementia-free life." Sister Marcella's "life, cognitive abilities and longevity were extraordinary." At age 100, her Mini-Mental Status Examination score was 28/30 indicating very good cognitive functioning. At postmortem, "she had a remarkably clean, large brain" when she died at 101.

Of 27 centenarian brains, there were three "supernormal" centenarians who had retained good mental functioning; in their neuropathological findings "no apparent senile changes or ischemic lesions" were found. The postmortem findings of the 27 centenarian brains were "not fundamentally different" than findings of brains of less elderly.[7]

Research showing a decreased rate of incidence of dementia by 13% per calendar decade in Europe and in the USA is also encouraging. Healthy aging without dementia is achievable[7–11] even with lesions postmortem.[7]

Neuroscience provides evidence-based interventions to "prevent, delay onset, and/or reverse" cognitive decline and dementia[12–14] and to enhance brain architecture and function.[15–30]

Studies suggest that senile dementia can be seen as "age-related" because it occurs within specific ages, "rather than as an 'age-related' disorder (that is, caused by the aging process itself)."[31] Dutch centenarians who scored 26 or above on the MMSE maintained their high level of cognitive performance two years later.[32,33]

Although percentages varied, there were centenarians with no dementia in many countries. The Fordham Centenarian Study found no, or very few, cognitive limitations in 119 centenarians.[10] A review of 11,084 electronic records of centenarians in the United Kingdom found only 12% of women and 6% of men diagnosed with dementia.[34] Almost 57% of 228 Chinese individuals aged 100–112 were independent.[35] The Framingham Heart Study[36] found a decline in incidence of dementia across three decades.

In 57 dementia-free centenarians, the finding that "stronger functional connectivity between right frontoparietal control network," and a stronger functional connectivity compared to subjects aged 76–79, demonstrated a more intact bilateral neuronal efficiency. This "may relate to the resistance to cognitive decline in our cohort of dementia-free near-centenarians and centenarians who can be considered as successful agers."[37]

One theory about how some remain cognitively efficient proposes that, during their prenatal and neonatal life, neuron selection for organizing specific areas of the cortex was more accurate. Thus, less neurons which were vulnerable to degeneration promoted by amyloid deposits were able to survive. This left a healthier and stronger array of neurons for cortical neural circuits which could resist age-related changes.[38]

A compression of morbidity has been documented.[39–42] Thriving more than a century in good health with compressed morbidity is "a good model of successful aging."[42]

Among the reasons exercise remains important is that it increases brain factors which might play a role in "reversing brain aging" and "improve cognitive functioning."[43–45] Chinese elders with MCI appreciated cognitive gains, less depression, and better balance after 12 weeks of square dancing.[46] After 18 months of dancing, participants had improved balance as well as increased volume of the hippocampus[47] as measured by MRI as compared to individuals whose physical fitness routine was conventional. Increased plasma BDNF was found in dancers whose attention and spatial memory were improved.[48] Much more on this in later chapters.

As the artist creating the artwork of our evolving brains, we are empowered by the research our neuroscientists provide. Laughter and celebration of our efforts to enhance the chemistry, architecture, and function of our plastic brains would be beneficial to us and potentially to those we spend time with as described above.

With our efforts, we can have FUN influencing Flynn findings further. We have a smorgasbord of choices to celebrate and enjoy in our efforts to prevent, delay onset, and/or reverse dementia by building bigger, better brains.

We are the artisans. Let the dance begin! Let our joy and laughter enhance our brain artworks-in-process as we celebrate that our brains are plastic.

References

1. Diamond MC. (1988). *Enriching Heredity: The Impact of the Environment on the Anatomy of the Brain*. New York: The Free Press.
2. Flynn JR. (2012). *Are We Getting Smarter? Rising IQ in the Twenty-First Century*. New York: Cambridge University Press.
3. Flynn JR. (2018). Reflections about Intelligence over 40 Years. *Intelligence*, 70: 73–83. https://doi.org/10.1016/j.intell.2018.06.007
4. Wongupparaj P, Wongupparaj R, Morris RG & Kumari V. (2023). Seventy Years, 1000 Samples, and 300,000 SPM Scores: A New Meta-Analysis of Flynn Effect Patterns. *Intelligence*, 98: 101750. https://doi.org/10.1016/j.intell.2023.101750
5. Gonot-Schoupinsky F. (2023) From Positive Psychology to Positive Biology: Laughter and Longevity. *Explor Med*, 4: 1109–1115. https://doi.org/10.37349/emed.2023.00198
6. Snowdon DA. (2003). Healthy Aging and Dementia: Findings from the Nun Study. *Ann Intern Med*, 139: 450–454. https://doi.org/10.7326/0003-4819-139-5_part_2-200309021-00014
7. Mizutani T & Shimada H. (1992). Neuropathological Background of Twenty-Seven Centenarian Brains. *J Neurol Sci*, 108: 168–177. https://doi.org/10.1016/0022-510X(92)90047-O

8. Andersen-Ranberg K, Vasegaard L & Jeune B. (2001). Dementia Is Not Inevitable: A Population-Based Study of Danish Centenarians. *J Gerontol B Psychol Sci Soc Sci*, 56: 152–159. https://doi.org/10.1093/geronb/56.3.P152

9. Perls TT. (2004). Centenarians Who Avoid Dementia. *Trends Neurosci*, 27: 633–636. https://doi.org/10.1016/j.tins.2004.07.012

10. Jopp DS, Park M-KS, Lehrfeld J & Paggi ME. (2016). Physical, Cognitive, Social and Mental Health in Near-Centenarians and Centenarians Living in New York City: Findings from the Fordham Centenarian Study. *BMC Geriatr*, 16: 1. https://doi.org/10.1186/s12877-015-0167-0

11. Qiu C & Fratiglioni L. (2018). Aging without Dementia Is Achievable: Current Evidence from Epidemiological Research. *J Alzheimer Dis*, 62: 933–942. https://doi.org/10.3233/JAD-171037

12. Ball KK, Berch DB, Helmers KF, Jobe JB, Leveck MD, Mariske M, Morris JN, Rebok GW, Smith DM, Tennstedt SL, Unverzagt FW, Willis SL, and the ACTIVE Study Group. (2002). Effects of Cognitive Training Interventions with Older Adults: A Randomized Controlled Trial. *JAMA*, 288: 2271–2281. https://doi.org/10.1001/jama.288.18.2271

13. Ball KK, Ross LA, Roth DL & Edwards JD. (2013). Speed of Processing Training in the Active Study: Who Benefits? *J Aging Health*, 25: 65S–84S. https://doi.org/10.1177/0898264312470167

14. Mahncke H, Bronstone A & Merzenich MM. (2006). Brain Plasticity and Functional Losses in the Aged: Scientific Bases for a Novel Intervention. *Prog Brain Res*, 157: 81–109. doi: 10.1016/S0079-6123(06)57006-2

15. Diamond MC. (1988). *Enriching Heredity: The Impact of the Environment on the Anatomy of the Brain*. New York: The Free Press.

16. Diamond MC. (2001). Response to the Brain of Enrichment. *An Acad Bras Cienc*, 73: 211–220. https://doi.org/10.1016/B0-08-043076-7/03626-3

17. Pereira AC, Huddleston DE, Brickman AM, Sosunov AA, Hen R, McKhann GM, et al. (2007). An in Vivo Correlate of Exercise-Induced Neurogenesis in the Adult Dentate Gyrus. *Proc Natl Acad Sci U S A*, 104: 5638–5643. https://doi.org/10.1073/pnas.0611721104

18. Angevaren M, Aufdemkampe G, Verhaar HJJ, Aleman A & Vanhees L. (2008). Physical Activity and Enhanced Fitness to Improve Cognitive Function in Older People without Known Cognitive Impairment. *Cochrane Database Syst Rev*, 2: CD005381. https://doi.org/10.1002/14651858.CD005381.pub3

19. Larson EB. (2008). Physical Activity for Older Adults at Risk for Alzheimer Disease. *JAMA*, 300: 1077–1079. https://doi.org/10.1001/jama.300.9.1077

20. Baker LD, Frank LL, Foster-Schubert K, Green PS, Wilkinson CW, McTiernan A, et al. (2010). Effects of Aerobic Exercise on Mild Cognitive Impairment: A Controlled Trial. *Arch Neurol*, 67: 71–79. https://doi.org/10.1001/archneurol.2009.307

21. Lojovich JM. (2010). The Relationship between Aerobic Exercise and Cognition: Is Movement Medicinal? *J Head Trauma Rehabil*, 25: 184–192. https://doi.org/10.1097/HTR.0b013e3181dc78cd

22. Erickson KI, Voss MW, Prakash RS, Basak C, Szabo A, Chaddock L, et al. (2011). Exercise Training Increases Size of Hippocampus and Improves Memory. *Proc Natl Acad Sci U S A*, 108: 3017–3022. https://doi.org/10.1073/pnas.1015950108

23. Erickson KI, Leckie RL & Weinstein AM. (2014). Physical Activity, Fitness, and Gray Matter Volume. *Neurobiol Aging*, 35: 20–28. https://doi.org/10.1016/j.neurobiolaging.2014.03.034

24. Jessberger S & Gage FH. (2014). Adult Neurogenesis: Bridging the Gap between Mice and Humans. *Trends Cell Biol*, 24: 558–563. https://doi.org/10.1016/j.tcb.2014.07.003

25. Nagamatsu LS, Flicker L, Kramer AF, Voss MW, Erickson KI, Hsu CL, et al. (2014). Exercise Is Medicine, for the Body and the Brain. *Br J Sports Med*, 48: 943–944. doi: 10.1136/bjsports-2013-093224. Epub 2014 Mar 21.

26. Niemann C, Godde B & Voelcker-Rehage C. (2016). Senior Dance Experience, Cognitive Performance, and Brain Volume in Older Women. Neural Plast: 9837321. https://doi.org/10.1155/2016/9837321

27. Ryan SM & Nolan Y. (2016). Neuroinflammation Negatively Affects Adult Hippocampal Neurogenesis and Cognition: Can Exercise Compensate? Neurosci *Biobehav Rev*, 61: 121–131. https://doi.org/10.1016/j.neubiorev.2015.12.004

28. Shaffer J. (2016). Neuroplasticity and Clinical Practice: Building Brain Power for Health. *Front Psychol*, 7: 1118. https://doi.org/10.3389/fpsyg.2016.01118

29. Burzynska AZ, Jiao Y, Knecht AM, Fanning J, Awick EA, Chen T, et al. (2017). White Matter Integrity Declined Over 6-Months, but Dance Intervention Improved Integrity of the Fornix of Older Adults. *Front Aging Neurosci*, 9: 59. https://doi.org/10.3389/fnagi.2017.00059

30. Edwards JD, Fausto BA, Tetlow AM, Crorna RT & Valdés EG. (2018). Systematic Review and Meta-Analyses of Useful Field of View Cognitive Training. *Neurosci Biobehav Rev*, 84: 72–91. https://doi.org/10.1016/j.neubiorev.2017.11.004

31. Ritchie K & Kildea D. (1995). Is Senile Dementia "age-related" or "ageing-related"? – Evidence from Meta-Analysis of Dementia Prevalence in the Oldest Old. *Lancet*, 346: 931–934. DOI: 10.1016/s0140-6736(95)91556-7

32. Beker N, Sikkes SAM, Hulsman M, Schmand B, Scheltens P & Holstege H. (2019). Neuropsychological Test Performance of Cognitively Healthy Centenarians: Normative Data from the Dutch 100-Plus Study. *J Am Geriatr Soc*, 67: 759–767. https://doi.org/10.1111/jgs.15729

33. Beker N, Sikkes SAM, Hulsman M, Tesi N, van der Lee SJ, Scheltens P, et al. (2020). Longitudinal Maintenance of Cognitive Health in Centenarians in the 100-Plus Study. *JAMA Netw Open*, 3: e200094. https://doi.org/10.1001/jamanetworkopen.2020.0094

34. Hazra NC, Dregan A, Jackson S & Gulliford MC. (2015). Differences in Health at Age 100 According to Sex: Population-Based Cohort Study of Centenarians Using Electronic Health Records. *J Am Geriatr Soc*, 63: 1331–1337. https://doi.org/10.1111/jgs.13484

35. Huang Z, Chen Y, Zhou W, Li X, Qin Q, Fei Y, et al. (2020). Analyzing Functional Status and Its Correlates in Chinese Centenarians: A Cross-Sectional Study. Nurs Health Sci: 1–9. https://doi.org/10.1111/nhs.12707. [Epub ahead of print].

36. Satizabal CL, Beiser AS, Chouraki V, Chene G, Dufouil C & Seshadri S. (2016). Incidence of Dementia over Three Decades in the Framingham Heart Study. *N Engl J Med*, 374: 523–532. https://doi.org/10.1056/NEJMoa1504327

37. Jiang J, Liu T, Crawford JD, Kochan NA, Brodaty H, Sachdev PS, et al. (2020). Stronger Bilateral Functional Connectivity of the Frontoparietal Control Network in Near-Centenarians and Centenarians without Dementia. *Neuroimage*, 215: 116855. https://doi.org/10.1016/j.neuroimage.2020.116855

38. Bugiani O. (2020). The Puzzle of Preserved Cognition in the Oldest Old. *Neurol Sci*, 41: 441–447. https://doi.org/10.1007/s10072-019-04111-y

39. Andersen SL, Sebastiani P, Dworkis DA, Feldman L & Perls TT. (2012). Health Span Approximates Life Span among Many Supercentenarians: Compression of Morbidity at the Approximate Limit of Life Span. *J Gerontol A Biol Sci Med Sci*, 67A: 395–405. https://doi.org/10.1093/gerona/glr223

40. Richmond RL, Law J & Kay-Lambkin F. (2012). Morbidity Profiles and Lifetime Health of Australian Centenarians. *Australas J Ageing*, 31: 227–232. https://doi.org/10.1111/j.1741-6612.2011.00570.x

41. Sebastiani P, Gurinovich A, Nygaard M, Sasaki T, Sweigart B, Bae H, et al. (2019). APOE Alleles and Extreme Human Longevity. *J Gerontol A Biol Sci Med Sci*, 74: 44–51. https://doi.org/10.1093/gerona/gly174

42. Hashimoto K, Kouno T, Ikawa T, Hayatsu N, Miyajima U, Yabukami H, et al. (2019). Single-Cell Transcriptomics Reveals Expansion of Cytotoxic CD4 T Cells in Supercentenarians. *Proc Natl Acad Sci U.S.A.*, 116: 24242–24251. https://doi.org/10.1073/pnas.1907883116

43. Baker LD, Bayer-Carter JL, Skinner J, Montine TJ, Cholerton BA, Callaghan M, et al. (2012). High-Intensity Physical Activity Modulates Diet Effects on Cerebrospinal β-Amyloid Levels in Normal Aging and Mild Cognitive Impairment. *J Alzheimer Dis*, 28: 137–146. https://doi.org/10.3233/JAD-2011-111076

44. Baker LD. (2015). Aerobic Exercise Reduces CSF Levels of Phosphorylated Tau in Older Adults with MCI. In *Alzheimer's Association International Conference 2015 Presentation* (Winston-Salem NC).

45. Horowitz AM, Fan X, Bieri G, Smith LK, Sanchez-Diaz CI, Schroer AB, et al. (2020). Blood Factors Transfer Beneficial Effects of Exercise on Neurogenesis and Cognition to the Aged Brain. *Science*, 369: 167–173. https://doi.org/10.1126/science.aaw2622

46. Wang S, Yin H, Meng X, Shang B, Meng Q, Zheng L, et al. (2019). Effects of Chinese Square Dancing on Older Adults with Mild Cognitive Impairment. Geriatr *Nurs*, 41: 290–296. https://doi.org/10.1016/j.gerinurse.2019.10.009

47. Rehfeld K, Müller P, Aye N, Schmicker M, Dordevic M, Kaufmann J, et al. (2017). Dancing or Fitness Sport? The Effects of Two Training Programs on Hippocampal Plasticity and Balance Abilities in Healthy Seniors. *Front Hum Neurosci*, 11: 305. https://doi.org/10.3389/fnhum.2017.00305

48. Rehfeld K, Luders A, Hoekelmann A, Lessmann V, Kaufmann J, Brigadski T, et al. (2018). Dance Training Is Superior to Repetitive Physical Exercise in Inducing Brain Plasticity in the Elderly. *PLoS ONE*, 13: e0196636. https://doi.org/10.1371/journal.pone.0196636

2

We Must Be Doing Some Things Right

The Flynn effect suggests we must be doing some things right since increases in human intelligence have been documented in recent decades in 34 countries.[1–3] Repeating that is related to how much FUN can be had in activities influencing further Flynn effect increases locally and around the world for *all* involved.

In my youth, I spent as much time as possible sitting on the floor at the feet of my loving grandmother watching her spend hours reading. That was among the positive influences on my choosing activities that influenced the evolving artwork of my very young brain. She was doing the right thing. She modeled approval of wanting to read.

We have many personal and documented inspiring examples to learn from. Before the word neuroplasticity entered our vocabulary, these role models were enhancing their neuroplasticity by doing some things right.

Positive Agers

Data classified 1303 participants in the UK Biobank into groups based on their fluid intelligence test results which had been collected over seven to nine years with a touch screen questionnaire.[4] The result was 77% accuracy in predicting "positive aging" versus cognitive decline. The 563 positive agers had a positive slope in their test results, indicating enhancing brain function over time; 360 maintained neurocognitive functioning with little change over time; and 380 participants had a decline over time in their fluid intelligence measures. Plasma samples were used to assess the metabolomics array.

DOI: 10.4324/9781003462354-2

Participants with positive aging had low-density lipoprotein (LDL) particles with a higher average diameter which were less likely to "contribute to plaque formation." Those with higher platelet counts were more likely to show neurocognitive decline which "may be due to increased risk of thrombosis that could adversely affect blood flow and nutrient transport to the brain." Levels of pyruvate were also higher at baseline in the positive-aging group. They opined that their "results suggest that optimal cognitive aging may not be related to age per se but biological factors that may be amenable to lifestyle or pharmacological changes." With this more positive perspective on healthy aging of our brains, research on the individuals who maintain and/or improve good neurocognitive skills at advanced ages could enhance successful aging for the general population. This research needs to "incorporate longitudinal structural, functional, and molecular brain imaging markers using magnetic resonance imaging (MRI), in order to better predict cognitive trajectories." This is one of many ways that artificial intelligence used with human oversight and creativity could expedite this essential research and hacking neuroplasticity.

In a longitudinal population-based study, people with an average age of 68.8 years performed a memory task while undergoing functional magnetic resonance imaging (fMRI) scans.[5] They "were classified as successful or average" on cognitive development during their 15–20 years in the study. There were 51 successful elders and 51 age-matched elders. The BOLD signal of their fMRI was higher in successful elders than in the average participants "notably in the bilateral PFC and left hippocampus." The hippocampal activation correlated with memory performance. In the average elders, this activation was less than in a reference group 35.3 years old but, in the successful elders, it was *not* less than the 35.3-year-old reference group. They opined that "one mechanism behind successful cognitive aging might be preservation of HC function combined with a high frontal responsivity." HC is hippocampus, the part of your brain that is associated with memory and learning. Human creativity and oversight of artificial intelligence used in developing wearables to assess these brain signals could add to our Intelligent Neuroplasticity Revolution so we can continue doing several things right.

SuperAgers

A growing body of research on "SuperAgers" can increase the resolve to responsibly employ artificial intelligence in using evolving neuroscience to learn from these role models of excellent brain health across extended lifespans, identify how to best assess and meet their needs, and to develop applications to enhance the Flynn effect more effectively, more broadly, and for more of our global citizens. Although the typical pattern has been for memory to decline with aging, our SuperAgers have memory functions like individuals 20 years younger than their age at assessment. Thus, SuperAgers afford us a new perspective about cognitive aging.

The tsunami of research reporting on SuperAgers is valuable to help us design our Intelligent Neuroplasticity Revolution. In reviewing 21 studies of these cognitively resilient adults, it was noted that the way that they preserved their cortex was selective.[6,7] "In this context, the anterior cingulate cortex is highlighted as an imaging and histologic signature of these subjects." Compared to age-matched control individuals, the anterior cingulate cortex was significantly thicker in SuperAgers as was their hippocampus. Their memory skills were positively correlated with the average cingulate cortex cortical thickness in these adult SuperAgers. It is worth noting that amyloid deposits were of similar levels in SuperAgers as in controls. These researchers opined that "brain resilience may be partially independent of neurodegeneration."

Humans whose episodic memory is the same as or better than middle-aged adults are considered SuperAgers who give us a positive role model and provide an opportunity to study how their brains achieve this excellent neuro-cognitive resilience.[8] Detailed neuropsychological testing included

> the Forward and Backward Digit Span, Trail-Making A and B, Verbal Fluency (animals) and Letter Verbal Fluency tests, Rey-Osterrieth Complex Figure (copy and delayed recall), Logical Memory of the Wechsler Memory Scale, Rey Auditory Verbal Learning Test, 60-item version of the Boston Naming Test, and Estimated Intelligence Quotient measured with the Wechsler Adult Intelligence Scale, Third Edition.

From a group of individuals 80 or more years old, they identified 14 who qualified as SuperAgers and a control group of 17 people who were cognitively average. Resting-state-fMRI at 3T and 7T MR imaging was obtained on these participants. "The key networks that differentiated superagers and elderly controls were the default mode, salience, and language networks." They opined that these functional connectivity findings "can provide potential imaging biomarkers for predicting superagers. The 7T field holds promise for the most appropriate study setting to accurately detect the functional connectivity patterns in superagers."

One study defined SuperAgers as being between the ages of 60–80 and performing "at or above the value for young adults" aged 18–32 years "on the Long Delay Free Recall measure of the" California Verbal Learning Test (CVLT) as well as "no lower than 1 SD below the mean for their age group on the Trail Making Test Part B".[9] SuperAgers

> performed better than typical older adults on CVLT Trial 1, Trial 5, total learning, Long Delay Free Recall intrusion rate, and Long Delay Free Recall semantic clustering, and showed a trend toward better performance on Long Delay Recognition memory accuracy.

SuperAgers had a significantly thicker cortex in some locations. They also had preserved the volume of their hippocampus which was associated with their memory performance. These researchers opined that their "results indicate older adults with youthful memory abilities have youthful brain regions in key paralimbic and limbic nodes of the default mode and salience networks that support attentional, executive, and mnemonic processes subserving memory function."

Another study considered SuperAgers to be over 80 years old "with episodic memory performance at least as good as normative values for 50- to 65-year-olds."[10] The cerebral cortex of these SuperAgers was significantly thicker than their healthy age-matched peers and displayed no atrophy compared to the younger healthy group. Also, "a region of left anterior cingulate cortex was significantly thicker in the SuperAgers than in both elderly and middle-aged controls." Artificial intelligence could assist in the development of essential research to understand factors related to how these SuperAgers had preserved cognitive and neuroanatomical benefits and resisted the so-called "normal age-related" decline in brain volume and memory capacity.

Another study did a "retrospective analysis of neuropsychological test data from SuperAgers" who were over 80 years old and had superior performance on episodic memory and "at least average-for-age performance in non-episodic memory domains."[11] These 56 SuperAgers were compared to 23 peers of a similar age who had average episodic memory. Compared to their average-memory peers, these SuperAgers had better scores "on measures of attention, working memory, naming, and speeded set-shifting." Variability in nonepisodic memory of these SuperAgers includes some being "above average-for-age across cognitive domains while others performed in the average-for-age range on non-memory tests." In all participants, over 20% of the difference in their scores on episodic memory was explained by their scores on executive function and attention.

Centenarian Super Successes

Among the centenarians to be inspired by is "Howard Tucker (USA, b. 10 July 1922) who is 98 years 231 days old, as verified in Cleveland, Ohio, USA, on 26 February 2021" when he was listed on the GuinnessWorldRecords. com as the oldest physician in practice.[12] CNBC.com includes an interview with this gentleman.[13]

He has been a "practicing doctor and neurologist for more than seven decades" and is still seeing patients at the age of 101. He emphasizes consistent brain stimulation which he demonstrates by staying on the job. Howard Tucker, MD, is an inspiring role model. He strengthens my resolve to keep the promise I've made to still be on the job at Harborview Medical Center when I am 150.

Our Intelligent Neuroplasticity Revolution needs to consider multiple evidence-based interventions and assessments in our efforts to hack neuroplasticity at *every* age. As our population ages around the globe, we can increase the extent to which the oldest old contribute to, as well as benefit from, the types and methods of interventions that artificial intelligence can expedite making age appropriate and culturally tailored.

Examples of contributions in addition to Dr. Tucker are already impressive given what we learn about through the media. These centenarian super successes are of increasing importance as role models.

Pulitzer Prize winning author Herman Wouk wrote his memoir as a centenarian entitled *Sailor and Fiddler: Reflections of a 100-Year-Old.*[14] He said he was a "happy gent" at age 100. He lived to be 103 years old.

Eileen Kramer in Australia continued to dance and be a choreographer at the age of 103.[15] As found on Wikipedia, she was born in 1914, began her studies in the 1930s, became a dancer, and in 2021, was still active in the arts.

Bob Vollmer made the news when he retired as a surveyor for the Indiana Department of Natural Resources at the age of 102.[16] They report that he attributed his longevity to parental care, including nutritional meals as well as his grandmother's advice to "Eat a lot of red beets."

Man Kaur started running at the age of 90 when her son coached her; she was still running in New Zealand at the age of 101. As found on Wikipedia,[17] by age 103, she earned "the world records in the Over-100 years old categories for a variety of events" in running.

Tillie Dybing survived COVID-19 at the age of 107 after having also experienced the 1918 Spanish flu.[18] She survived two pandemics.

Since I was one of 247 bicyclists who signed up for the Odyssey 2000© bicycle ride around the world with the goal of bicycling 80 miles a day, it's *special* to share the record of Robert Marchand[19] accessed 10 February 2024. He set three world records for track bicycling. After his first world record at age 100, he attracted the attention of researchers who guided him on a training program. Marchand set his second world record at age 103; he was 11% faster at age 103 than he was at 100.

Perhaps most important about his performance improvements was that Marchand gave us metric evidence of improving cardiovascular functioning in a centenarian, the *first* such evidence in a centenarian!! His was a 13% improvement in cardiovascular functioning at age 103!! At the age of 105, he set his third world record and, when it did not exceed his previous records, you must love his attitude: it is reported that he said: "If I had seen the 10-minute warning card, I would have pedaled slightly faster." He was still bicycling at the age of 107.

Supercentenarian Super Stars

Healthy aging without dementia is achievable in centenarians and even in supercentenarians, individuals aged 110 or more. This is illustrated in a brief communication about a woman who lived to age 115 years.[20] "In 2001, at the age of 111, she inquired whether her frail body, after death, was still useful for scientific research or teaching purposes." Her poor eyesight limited tests that could be used. The MMSE was done without the visual items; her attention was measured with the WAIS Digits Forward; her working memory was measured with WAIS Digits Backwards and Serial Sevens; and verbal reasoning used a similarities subtest. Her mental arithmetic was "Errorless and fast except division." At the age of 113 years and 10 months, she earned scores which were "still within the normal range for healthy older adults" of 60–76 years of age. After her death at 115, there was "no significant atherosclerosis, and her brain had no vascular pathology, only a slight amount of tau pathology." Another *very* inspiring role model is described here!!

Examples of extraordinary cognitive and physical health in supercentenarians includes a woman 118 years and 9 months old whose "language skills were largely preserved" and whose memory "was withing the normal limits for persons over 80 years" despite impaired hearing and vision.[21] This highlights the urgency of the need to develop measurements, norms, and strategic use of evidence-based interventions to maximize driving neuroplasticity in a positive direction for our oldest old.

"The Myth of Cognitive Decline"

For years, I speculated and taught this with the analogy of handling accumulating paperwork. In the early decades of my life, every very important document could easily be carried in a briefcase. Over the years, that paperwork expanded into 27 drawers of large metal filing cabinets. When a specific document was needed, I had to remember which filing cabinet to go to, which drawer to open, which file to pull out, and then sort through the many papers in that file. That cycle was repeated until the vital document was found.

"The Myth of Cognitive Decline" is a peer-reviewed article written by Ramscar and colleagues describing their existing and ongoing research.[22] While they note that there are changes in performance on many psychometric tasks with aging, they opine that their long series of studies indicates that these changes reflect the "consequences of learning on information processing, and not cognitive decline." Across the years, adults develop a greater sensitivity to the small details of differences in stimuli, accumulate more acquired knowledge, and, as a result, have more and different demands in their memory search "which escalate as experience grows." These researchers concluded that the performance of older adults reflects predictable outcomes of increased learning on information processing; it is *not* an indication of cognitive decline; it is a call for different search engine processes.

That's a much more elegant way to explain how to access accumulated learning. It is also based on years of research. Their article "The Myth of Cognitive Decline" was simultaneously a mood elevator as well as a time saver. Surely, that perspective is even more pertinent now that real-time access to vast amounts of data 24/7 can result in many of us being on cognitive overload. More learning can help our *Enriching Heredity* and give us additional reasons for celebrating how much artificial intelligence can assist us with that.

Positive Empowering Perspectives

All of this indicates that we are doing some things right. Accumulating more knowledge in our healthy-aging brains enhances neuroplasticity. Our searches and coordination of data to add to this memory storage can be expedited with artificial intelligence. Viewing cognitive differences across the lifespan as related to better awareness of details and accumulated learning is an example of a hopeful perspective which could effectively empower increasing efforts to enhance healthy neuroplasticity across the lifespan; this could be good for business, education, career, health, and prosocial behaviors.

References

1. Flynn JR. (2012). Are We Getting Smarter? Rising IQ in the Twenty-First Century. New York: Cambridge University Press.
2. Flynn JR. (1987). Massive IQ Gains in 14 Nations: What IQ Tests Really Measure. *Psychol Bull*, 101: 171–191.
3. Flynn JR. (2018). Reflections about Intelligence Over 40 Years. *Intelligence*, 70: 73–83. https://doi.org/10.1016/j.intell.2018.06.007
4. Mohammadiarvejeh P, Klinedinst BS, Wang Q, Li T, Larsen B, Pollpeter A, Moody SN, Willette SA, Mochel JP, Allenspach K & Hu G. (2023). Bioenergetic and Vascular Predictors of Potential Super-Ager and Cognitive Decline Trajectories—A UK Biobank Random Forest Classification Study. GeroScience, 45: 491–505. https://doi.org/10.1007/s11357-022-00657-6
5. Pudas S, Persson J, Josefsson M, de Luna X, Nilsson L-G & Nyberg L. (2013). Brain Characteristics of Individuals Resisting Age-Related Cognitive Decline over Two Decades. *J Neurosc*, 33(20): 8668–8677. https://doi.org/10.1523/JNEUROSCI.2900-12.2013
6. de Godoy LL, Alves CAPF, Saavedra JSM, Studart-Neto A, Nitrini R, Leite Cd C & Bisdas S. (2021) Understanding Brain Resilience in Superagers: A Systematic Review. *Neuroradiology*, 63: 663–683. https://doi.org/10.1007/s00234-020-02562-1
7. de Godoy LL, Studart-Neto A, Wylezinska-Arridge M, Tsunemi MH, Moraes NC, Yassuda MS, Coutinho AM, et al. (2021) The Brain Metabolic Signature in Superagers Using In Vivo 1H-MRS: A Pilot Study. *AJNR Am J Neuroradiol*, 42(10): 1790–1797. https://doi.org/10.3174/ajnr.A7262
8. de Godoy LL, Studart-Neto A, de Paula DR, Green N, Halder A, Arantes P, Chaim KT, et al. (2023) Phenotyping Superagers Using Resting-State fMRI. *Am J Neuroradiol*, 44(4): 424–433. https://www.ajnr.org/content/early/2023/03/16/ajnr.A7820
9. Sun FW, Stepanovic MR, Andreano J, Barrett LF, Touroutoglou A, & Dickerson BC. (2016). Youthful Brains in Older Adults: Preserved Neuroanatomy in the Default Mode and Salience Networks Contributes to Youthful Memory in Superaging. *J Neuroscience*, 36(37): 9659–9668.
10. Harrison TM, Weintraub S, Mesulam MM & Rogalski E. (2012). Superior Memory and Higher Cortical Volumes in Unusually Successful Cognitive Aging. *J Int Neuropsychol Sci*, 18(6): 1081–1085. https://doi.org/10.1017/S1355617712000847

11. Maher AC, Makowski-Voidan B, Kuang A, Zhang H, Weintraub S, Mesulam MM & Rogalski E. (2022). Neuropsychological Profiles of Older Adults with Superior Versus Average Episodic Memory: The Northwestern 'SuperAger' Cohort. *J Int Neuropsychol Soc*, 28(6): 563–573. https://doi.org/10.1017/S1355617721000837
12. Oldest doctor (male) | Guinness World Records 26 February 2021.
13. At 101 Years Old, I'm the 'world's oldest practicing doctor': My No. 1 Rule for Keeping Your Brain Sharp (cnbc.com)
14. Wouk H. (2016). Sailor and Fiddler: Reflections of a 100-Year-Old Author. New York: Simon & Schuster.
15. Eileen Kramer - Wikipedia
16. Scipioni J. (2020) At 102, Indiana's oldest state employee says this is the secret to his long success. 102-year-old Bob Vollmer says secret to long success is reading (cnbc.com).
17. Man Kaur - Wikipedia
18. Tillie Dybing, 107, Is Survivor Of Both COVID And 1918 Pandemic: 'I Feel Fine' - CBS Minnesota (cbsnews.com)
19. Ebbers C. (2017) What We Can Learn from This 105-Year-Old Cyclist. What We Can Learn from This 105-Year-Old Cyclist (outsideonline.com).
20. den Dunnen WFA, Brouwer WH, Bijlard E, Kamphuis J, van Linschoten K, Eggens-Meijer E, et al. (2008). No Disease in the Brain of a 115-Year-Old Woman. *Neurobiol Aging*, 29: 1127–1132. https://doi.org/10.1016/j.neurobiolaging.2008.04.010
21. Ritchie K. (1995) Mental Status Examination of an Exceptional Case of Longevity. J. C. Aged 118. *Br J Psychia*, 166: 229–235.
22. Ramscar M, Hendrix P, Shaoul C, Milin P & Baayen H. (2014). The Myth of Cognitive Decline: Non-Linear Dynamics of Lifelong Learning. Top Cogn Sci, 1–38. ISSN:1756-8757 print/1756-8765 online. https://doi.org/10.1111/tops.12078

3

Intelligent Neuroplasticity Revolution Empowers You

Gifts of Real-Time Data Access

We are gifted to be alive in the age of technology. This gives us access to evolving research in the field of our interest even before it goes to the press.

We are gifted to be alive when artificial intelligence is part of this rapidly changing technology. More than ever we now have the capacity to use our human reasoning, judgment, imagination, creativity, and problem solving to influence the use of artificial intelligence for analysis of variance, pattern recognition, image analysis, automation, and information processing.[1]

Recent examples of this include the development of a smart phone application using a ratiometric fluorescent technique to assess the amount of melatonin in a human.[2] Melatonin is a crucial neurohormone which our body produces. The multiple functions as described herein include sleep regulation and managing disease.[3]

Skin autofluorescence is a noninvasive way to estimate the accumulation of advanced glycation end products (AGEs) as an estimate of risk for dementia.[4] Your body's proteins react to sugar molecules that you consume by decreasing protein function and reducing elasticity in your skin, tendons, and blood vessels through irreversible rearrangements creating advanced glycation end products (AGEs). Thus, the device developed for measuring the accumulation of AGEs could help influence healthier dietary behaviors.

With good human reasoning, judgment, imagination, and creativity, we can use artificial intelligence to harvest real-time neuroscience data and

DOI: 10.4324/9781003462354-3

neurobiological status to hack our neuroplasticity for our best possible brain chemistry, architecture, and performance in ways that avoid accumulating the damage of AGEs. This can expedite the sharing of this important information to maximize the number of lives that could benefit. Our Intelligent Neuroplasticity Revolution empowers us, is timely, is urgent, and can be shared with our citizens around the globe.

I have been intrigued by the broad theme and variation in the human condition since my youth as number seven of seven born in ten years on a farm where we lived in poverty. Farms around us were prosperous. Then, as a Peace Corps dependent living on Borneo Island and traveling in that vicinity, it became painfully obvious that I knew *nothing* about poverty in comparison to what many of our world citizens struggle through. I returned to the USA with the new and strong mission to empower the gifted who know how to make our world a better place for *all*. We can maximize the benefits of our gifts of real-time access by increasing the *use* of evolving neuroscience for "*Enriching Heredity.*" [5] We welcome you to join in using creativity, imagination, and good judgment with the aid of artificial intelligence to capture the gifts of real-time data access to facilitate and expedite our Intelligent Neuroplasticity Revolution which empowers you to build bigger, better brains.

Tech Talks Together to Tailor

That perspective is the goal of this writing of an extensive overview of evolving neuroscience. Nobody can read the entire library, not even in their own area of specialization. Thus, this overview is intended to bring additional information to the current knowledge of several groups.

Government officials can heed the request of researchers referenced herein who are saying that this science needs to be reflected in government policy. Corporate decision makers can apply these principles in business settings to enhance brain power, increase happiness, promote health, and increase productivity of all people they influence as a proactive and mutually beneficial ways which could increase corporate impact while also being likely to reduce healthcare costs. Academicians can teach this evolving neuroscience and influence their students to carry it forward within their chosen career settings. Clinicians can make this an emphasis in the assessments and in care they provide from in utero through end of life. Philanthropic leaders can use this information to expedite the use of this data, which may need to include providing technology and devices to those who don't have them to expand the reach of these benefits. Communities can facilitate group activities organized around these evidence-based interventions to have healthy influences and prosocial benefits of increasing healthy aging and building positive communities for peaceful coexistence. Individuals can have tech talks together with their healthcare providers to tailor this technology to their specific needs, goals, and wishes.

All groups can (1) make copies of this available in their lending libraries to increase their strong positive influence on driving neuroplasticity in a positive direction in response to internal and external stimuli for the best possible brain health *and* brain power for *all;* (2) schedule regular tech talks together for discussions of (a) their understanding of the evidence-based interventions described within and (b) progress of bringing it into their lives and the lives of individuals that they influence; AND (3) have this influence in ways that also improve physical, mental, emotional, social, and neurobiological health for *all* global citizens.

Teamwork makes the team work to make the World a better place for *all.* Tech talks together can tailor the strategic activities and goals for your group.

Since the audience is broad, much of what is written herein will be highly scientific. Also, there will be words and examples without the scientific lexicon so that the nonscientists whom we seek to serve will enjoy getting the gist of what evolving neuroscience indicates could be of benefit for healthy aging for all who work with their healthcare provider in directing the focus of artificial intelligence to personalize their strategic path to their best brain chemistry, architecture, and performance.

You Have to Love our Neuroplasticity Progress

It is certainly encouraging that Flynn and colleagues documented increasing human intelligence during the 20th century in 34 countries.[6,7] This phenomenon is known as the Flynn effect. Does the Flynn effect mean we've been doing some things right? A study suggested that, despite having better cognitive functioning, how cognitive decline proceeds in the elderly is like what was previously observed.[8]

Additional reasons to love our progress in driving neuroplasticity in a positive direction are described in studies cited herein on factors that have been associated with an increase in the IQ scores and intelligence of individuals. We must be doing some things right.

Brookmeyer and colleagues[9] predicted that, if all we do is delay the onset of dementia by one year, we could have 9.2 million fewer cases by 2050. Celebrating while also doing everything possible to prevent dementia could increase the flow of neurochemicals associated with positive emotions; that is important because these neurochemicals of positive emotions enhance brain health. That's one of many reasons why this book focuses on providing hope for many courses of action to improve health while researchers work toward their goal of ending dementia. These research efforts to end dementia are essential but will take time to accomplish. Meanwhile, the estimate of Brookmeyer[9] and many subsequent studies can be additional wind needed beneath the wings of all who treasure and build brain health.

Another positive reframe is provided by Ramscar and colleagues[10] who wrote "The Myth of Cognitive Decline" in response to existing and ongoing research. Across the years, we develop a greater sensitivity to the small details of differences in stimuli, increase data stored in our neurons, and put more and different demands on our memory searches. The performance of older adults reflects predictable outcomes of increased learning on information processing. It is not an indication of cognitive decline. It is a call for different search engine processes.

Our memory searches and coordination of massive neuroscience data that is now available in real time can be expedited with artificial intelligence. Viewing cognitive differences across the lifespan as related to better awareness of details and accumulated learning is an example of a hopeful perspective which could effectively empower increasing efforts to enhance healthy neuroplasticity across the lifespan; this could be good for business, health, and prosocial behaviors. This is another reason to love our progress in neuroplasticity research and applications. It fuels my passion for reading, writing, teaching, and living these evidence-based techniques.

We can also be encouraged by the conclusion of The Lancet Commission on Dementia Prevention, Intervention and Care.[11] They reported that "dementia is by no means an inevitable consequence" of aging and that interventions across the lifespan might "delay or prevent a third of dementia cases."

A decline in the incidence and prevalence of dementia has been found across three decades in individuals who graduated from high school and "clinico-neuropathological studies found that nearly half of centenarians with dementia did not have sufficient brain pathology to explain their cognitive symptoms, while intermediate-to-high Alzheimer pathology was present in around one-third of very old people without dementia or cognitive impairment. This suggests that certain compensatory mechanisms (e.g., cognitive reserve or resilience) may play a role in helping people in extreme" old ages escape dementia syndrome.[12] This finding is another reason to love our neuroplasticity progress.

So, too, is the finding that participants in the Framingham Heart Study showed a decline in the incidence of dementia over three decades in countries with high income.[13] Since 1975, this analysis to determine dementia incidence in four 5-year epochs has included 5,205 people 60 years of age or older.

> Relative to the incidence during the first epoch, the incidence declined by 22%, 38%, and 44% during the second, third, and fourth epochs, respectively. This risk reduction was observed only among persons who had at least a high school diploma.

Despite a prevalence of vascular risk factors, their research did not completely clarify which factors contributed to this reduced incidence of dementia over three decades. However, it does argue in favor of continuing education and assuring education for all.

Love of our progress in driving neuroplasticity in a positive direction can include celebration of personal examples. We can watch it happen.

Joni Mitchell is a touching example. The Fortune[14] online article indicates that she learned to walk as a toddler, again at age nine when post-polio syndrome had her in a wheelchair, and again in 2015 after her brain aneurysm was nearly fatal. Her 2024 GRAMMYs[15] singing of "Both Sides Now" at 80 years of age and after surviving the aneurysm in 2015 is a very heartwarming personal example of why we can love the progress in neuroplasticity.

I've watched that video about ten times so far and will again for the love of Joni Mitchell's brilliance and inspirational use of driving neuroplasticity in a positive direction to heal in her youth and again as an adult to the extent that she made such an impressive comeback. Joni Mitchell is a model who inspires love for our progress in neuroplasticity. You probably have many examples in your friends and family who have also benefited from evidence-based interventions that could improve their neuroplasticity in inspiring ways.

FUN for Furthering Flynn[6,7]

The Flynn effect suggests that we must be doing some things right since increases in human intelligence have been documented in recent decades in

34 countries. Repeating that is related to how much FUN influencing further Flynn effect increases around the world could be for *all* involved.

We have many personal and documented inspiring examples to learn from. Before the word neuroplasticity entered our vocabulary, they were enhancing theirs by doing some things right. As described above, Joni Mitchell is one of those. As she beamed with happiness in her return to the stage, we can share in fun that could have positive effects on the chemistry, architecture, and performance of our brains.

Music making emerged to aid socialization and survival of humans through cooperation.[16] The ways that music can drive neuroplasticity in a positive direction are empowering in so many ways that music research will be included in several chapters of this writing.

AI for Our Intelligent Neuroplasticity Revolution

Using artificial intelligence with creative human imagination and judgment could influence and empower our Intelligent Healthy Neuroplasticity Revolution to *hasten* positive influence on the Flynn[6,7] effect of *increasing human intelligence*; aim to reduce cognitive decline with healthy aging; and plan to bring these neurocognitive benefits to many more of our global citizens as an essential effort to prevent, delay onset, and/or reverse dementia in ways that could simultaneously improve social, emotional, neurocognitive, physical, biological, immunochemical, and quality of life measures. AI can hasten the search of evidence-based interventions for driving neuroplasticity in a positive direction and the tailoring of them to the person and situation.

An example of this is described in a review of studies that included 14,966 people showing that healthier eating behaviors can be promoted with smartphone apps.[17] Smartphone apps health benefits can be possible in "an unsupervised, remote, and naturalistic setting in a large, community-based population." [18]

LOVE for the HEALTH of It

To love or not to love? There is no question!!

ALL our efforts begin with LOVE, in part because love of self will increase your likelihood of affording yourself the very best of care; also, because love and appreciation of others is so freely and easily shared.

As I walk the hallways of Harborview Medical Center, there are many people that can be thanked in passing by. "Thank you for making house beautiful," I say to any of the housekeepers without whom our facility could not stay as clean, safe, and healthy. Initially, this only puzzles some of them because they have become so accustomed to not being seen. With time, they return the smile. "Thank you for the excellent progress note you wrote on the patient in room 7 which markedly contributes to providing appropriate care" said to a clinician brightens their day as well as highlighting "appropriate care." Healthcare providers also require support and positive reinforcement of work well done. Love and using adversity for growth is constantly essential for healthy brain aging.

The second reason for this beginning with Love is that positive emotions are free and easy to share. After reading the research of Marian Diamond[5] presented herein about how powerful "TLC" is in building bigger, better brains, even a longer life, her research will likely strengthen your use and appreciation of "TLC." We can enhance the brain health, general health, and well-being of self and others as easily as sharing TLC.

Gratitude comes easily. It is among the many positive emotions that have been associated with giving your brain healthy chemistry that can protect and enhance your brain cells. Positive Psychology has given us much research to show the benefits of these positive emotions which are contagious.

That means that your own positive emotions could benefit those around you as well as yourself. Some say, and research cited herein confirms, it is not possible to give without receiving in return. With positive emotions, that is a good thing since our positive emotions can contribute to physical, mental, emotional, cognitive, career, business, and brain health for ALL. Research of Marian Diamond[4] cited herein emphasizes that "TLC" is an essential part of our Intelligent Neuroplasticity Revolution. LOVE for the HEALTH of it for ALL.

Hacking neuroplasticity with responsible use of artificial intelligence to personalize our Intelligent Neuroplasticity Revolution for enhancing healthy brain power is urgent. The UN/WHO Decade of Healthy Ageing[19] must include a focus on enhancing brain power in ways that include increasing prosocial behaviors as well as other components of health as will be made clear in studies referenced herein.

These interventions must continue far beyond the Decade of Healthy Ageing and become the new modus operandi. The purpose of this book is to give an extensive overview of the growing body of research suggesting how this can be influenced in a positive direction across the entire lifespan to influence a healthspan that approximates the extended lifespan.

Neuroscience provides evidence-based interventions to "prevent, delay onset, and/or reverse" cognitive decline and dementia.[20,21] This writing on evolving neuroscience emphasizes a lifespan focus. With advances in the age of technology, it is now possible to capture evidence noninvasively from the first moments of life. This influence can begin in utero and continue across the entire lifespan, including individuals with dementia and others with neurocognitive disorders.

Music heard prior to birth provides sufficient neuronal response to sustain fetal memory into early infancy.[22,23] Neonatal ICU (NICU) advances in care and technology have resulted in more premature babies surviving. Studies cited herein will highlight ways that time needing NICU level of care has been reduced; early intervention is essential to protect the neurodevelopment of these neonates and avoid neurodevelopmental disabilities such as visual impairment, cerebral palsy, and delayed socialization.

This is another time of life that artificial intelligence can expedite how we design research techniques to (1) maximize data on how to reduce the time premature infants need to be cared for in NICU, (2) clarify the types of care that maximize neuroplastic purchase, and (3) maintain maximum physical as well as neurocognitive health in neonates. Studies cited herein provide some thoughts on where to begin this use of artificial intelligence with human creativity and influence to help hack neuroplasticity at the beginning of life.

Music making emerged to aid socialization and survival of humans through cooperation.[16] The ways that music can drive neuroplasticity in a positive direction across our lifespan are empowering in so many ways that music research will be included in several chapters of this writing.

As centenarians and supercentenarians increase in number, many of them provide models of successful aging that can empower us as we learn from the wisdom of the ages. Even while our most well thought out and designed strategies for today empower us, we will constantly want to remember that it is a gift to have neuroscience evolve and bring us evidence that humans can use artificial intelligence responsibly in refining our protocols for our Intelligent Neuroplasticity Revolution going forward. Thus, we will make it a point to constantly revisit, update, and refine the multiple ways we will influence *"Enriching Heredity"*[4] for our unique version of ideal aging as we also celebrate the power of artificial intelligence to expedite our ready access to and use of the tsunami of evolving neuroscience.

Our Leverage for Furthering Flynn

ALL LANDS: **L**ove, **A**erobics, **N**eurobics, **D**iet, **S**leep. Using ALL of these categories has the potential Leverage for Furthering Flynn.

LOVE for the HEALTH of it has been increasingly important during the COVID pandemic. The struggles faced by so much of the world during the long confinement increased loneliness and included many lifestyle choices that increased mental health difficulties.[27]

Another reason this discussion, and the first chapter of this book, begins with **LOVE** for the HEALTH of it is that you are the only person in your universe who will spend twenty-four/seven with yourself. Thoughts are a choice. When we appreciate the many ways that our positive emotions can enhance our brain chemistry, architecture, and performance, we can be more highly motivated to frame things in the positive for the benefit of our brain health and the brain health of all around us. During this pandemic and civic unrest, each of us has more than the usual amount of adversity of some form and magnitude. So much so that it can become necessary to reframe it by saying something like: "This is another opportunity to expand my skills" in whatever way is essential at the time. We are seriously fortunate to have Diamond's findings[4] of the *huge and multiple* neuroplastic purchases of "TLC" described herein.

That's a good lead into why the second focus is on **A**erobics, Balance, Strength, AND Flexibility because Diamond's research was done with rats who had access to a running wheel.[4] Rats naturally run *many* meters on the running wheel daily. While we do not have data on whether they ran at an aerobic pace, stretched, or otherwise, studies herein will flesh out (pun intended) the serious impact of aerobic and strengthening exercise on increasing neurogenesis, increasing the chemistry of positive emotions, and improving the health of your brain cells. It is empowering to learn that "simply caloric expenditure, regardless of type or duration of exercise, may alone moderate neurodegeneration and even increase GM volume in structures of the brain central to cognitive functioning."[25] GM is the grey brain matter where your brain cells are. Increases in brain volume indicate that moderate to vigorous exercise has the potential to be neuroprotective.[26]

Neurobics, complex new learning, is next because we want to increase the survival rate of these new brain cells that we earned by being aerobic and doing resistance exercises. The new neurons are young and excitable; they require an invitation to join your complex brain; and the complex new learning you give them will influence the brain site where they will integrate.

Diet research for enhancing healthy brain and overall physical aging is sufficient to write a book that could be too heavy to carry. However, there's enough theme and variation in this research database to make it easier to eat for the health of it. That would include foods that reduce inflammation and oxidative stress as described herein to protect neurodevelopment.[29] Since energy consumed and energy used influence the rate of autophagy, research described herein will flesh out the reasons autophagy might become our favorite four-syllable word. The therapeutic effects of autophagy described in greater detail herein can increase brain cell longevity and reduce the aging of brain cells.[24]

> If we could give every individual the right amount of nourishment and exercise, not too little and not too much, we would have found the safest way to health.
>
> *Hippocrates*

Sleep. Deep sleep for slow wave sleep. Another enjoyable part of research cited herein is the focus on how *much* work and how *very* important body work gets done through our glymphatic system with deep quality and quantity of sleep. The recent American Heart Association scientific statement emphasizes the crucial role adequate quality and quantity of sleep plays in maintaining brain health.[28] The glymphatic system works best while we sleep.

In essence, *ALL* **L**ove, **A**erobics and other exercises, **N**eurobics for the joy of new learning, **D**iet components for healthy aging, and **S**leep for the health of some of your easiest brain and body work, *ALL* of these are part of the rich multiple-choice list of evidence-based interventions we can factor into how we move forward in hacking our neuroplasticity for our best possible healthy brain aging. Teamwork makes the team work. Thus, studies cited herein will expand on the added benefit of using more than one of these factors in each session. We will confer with our healthcare provider about our preferences and then use any available devices powered by artificial intelligence to set strategic goals, personalize them, culturally tailor them, prompt us when we forget something, monitor our progress, and give positive reinforcement for doing the things that can hack our neuroplasticity to improve our brain chemistry, architecture, and performance for the joy of our Intelligent Neuroplasticity Revolution.

References

1. Lawry T. (2023). *Hacking Health Care: How AI and the Intelligence Revolution Will Reboot an Ailing System*. New York: Routledge.

2. Megha KB, Arathi A, Shikha S, Alka R, Ramya P & Mohanan PV. (2024) Significance of Melatonin in the Regulation of Circadian Rhythms and Disease Management. *Mol Neurobiol*, 61(8): 5541–5571. https://doi.org/10.1007/s12035-024-03915-0. Epub 2024 Jan 11.
3. Kumar H & Obrai S. (2024) Ratiometric Fluorescent Sensing of Melatonin Based on Inner Filter Effect and Smartphone Established Detection. *Spectrochim Acta a Mol Biomol Spectrosc*, 304: 123309. https://doi.org/10.1016/j.saa.2023.123309.
4. Mooldijk SS, Lu T, Waqas K, Chen J, Vernooij, Ikram MK, Zillikens MC & Ikram MA. (2024) Skin Autofluorescence, Reflecting Accumulation of Advanced Glycation End Products, and the Risk of Dementia in a Population-Based Cohort. *Sci Rep*, 14(1): 1256. https://doi.org/10.1038/s41598-024-51703-6
5. Diamond MC. (1988). *Enriching Heredity: The Impact of the Environment on the Anatomy of the Brain*. New York: The Free Press.
6. Flynn JR. (2012). *Are We Getting Smarter? Rising IQ in the Twenty-First Century*. New Your: Cambridge University Press.
7. Flynn JR. (2018) Reflections about Intelligence over 40 Years. *Intel*, 70: 73–83. https://doi.org/10.1016/j.intell.2018.06.007
8. Gerstorf D, Ram N, Drewelies J, Duezel S, Eibich P, et al. (2022). Today's Older Adults Are Cognitively Fitter than Older Adults Were 20 Years Ago, but When and How They Decline Is No Different Than in the Past. *Psychol Sci*, 1–13. https://doi.org/10.1177/09567976221118541
9. Brookmeyer R, Johnson E. Ziegler-Graham K & Arrighi, M. (2007) Forecasting the Global Burden of Alzheimer's Disease. *Alzheimer's & Dementia*, 3: 186–191.
10. Ramscar M, Hendrix P, Shaoul C, Milin, P & Baayen H. (2014). The Myth of Cognitive Decline: Non-Linear Dynamics of Lifelong Learning. *Top Cogn Sci*, 6: 5–42. https://doi.org/10.1111/tops.12078
11. Livingston G, Huntley J, Sommerlad A, et al. (2020) Dementia Prevention, Intervention, and Care: 2020 Report of the Lancet Commission. *The Lancet*, 396: 413–436. https://doi.org/10.1016/S0140-6736(20)30367-6
12. Qiu C & Fratiglioni L. (2018). Aging without Dementia Is Achievable: Current Evidence from Epidemiological Research. *J Alzheimer Dis*, 62: 933–942. https://doi.org/10.3233/JAD-171037
13. Satizabal CL, Beiser AS, Chouraki V, Chene G, Dufouil C & Seshadri, S. (2016). Incidence of Dementia over three Decades in the Framingham Heart Study. *N Engl J Med*, 374: 523–532. https://doi.org/10.1056/NEJMoa1504327
14. After a Two-Decade Hiatus, Joni Mitchell Emerged as the Queen of Songstresses in the Best Year Yet for Women Artists | Fortune.
15. Watch: Joni Mitchell Performs "Both Sides Now" With Brandi Carlile, Allison Russell, Lucius, Jacob Collier, Blake Mills, and SistaStrings | 2024 GRAMMYs Performance | GRAMMY.com
16. Fukui H & Toyoshima K (2023) Testosterone, Oxytocin and Co-Operation: A Hypothesis for the Origin and Function of Music. *Front Psychol*, 14: 1055827. https://doi.org/10.3389/fpsyg.2023.1055827
17. Seid A, Fufa DD & Bitew ZW. (2024) The Use of Internet-Based Smartphone Apps Consistently Improved Consumers' Healthy Eating Behaviors: A Systematic Review of Randomized Controlled Trials. *Front Digit Health*, 6: 1282570. https://doi.org/10.3389/fdgth.2024.1282570

18. Sunderaraman P, Anda-Duran ID, Karjadi C, Peterson J, King H, Devine SA, Shih LC, Popp Z, Low S, Hwang PH, Goyal K, Hathaway L, Monteverde J, Lin H, Kolachalama VB & Au R. (2024) Design and Feasibility Analysis of a Smartphone-Based Digital Cognitive Assessment Study in the Framingham Heart Study. *J Am Heart Assoc*, 13(2): e031348. https://doi.org/10.1161/JAHA.123.031348.

19. UN Decade of Healthy Ageing – The Platform obtained 14 November 2023.

20. Ball KK, Berch DB, Helmers KF, Jobe JB, Leveck MD, Mariske M, Morris JN, Rebok GW, Smith DM, Tennstedt SL, Unverzagt FW, Willis SL and the ACTIVE Study Group. (2002). Effects of Cognitive Training Interventions with Older Adults: A Randomized Controlled Trial. *JAMA*, 288: 2271–2281. https://doi.org/10.1001/jama.288.18.2271

21. Mahncke H, Bronstone A & Merzenich MM. (2006). Brain Plasticity and Functional Losses in the Aged: Scientific Bases for a Novel Intervention. *Prog Brain Res*, 157: 81–109. doi: 10.1016/S0079-6123(06)57006-2.

22. Movalled K, Sani A, Nikniaz L & Ghojazadeh M. (2023). The Impact of Sound Stimulations during Pregnancy on Fetal Learning: A Systematic Review. BMC Pediatr, 23: 183. https://doi.org/10.1186/s12887-023-03990-7

23. Partanen E, Kujala T, Tervaniemi M & Huotilainen M. (2013). Prenatal Music Exposure Induces Long-Term Neural Efects. *PLoS ONE*, 8(10): e78946.

24. Marzoog BA. (2024). Autophagy as an Anti-Senescent in Aging Neurocytes. *Curr Mol Med*, 24(2): 182–190. https://doi.org/10.2174/156652402366623012010271820102718

25. Raji CA, Merrill DA, Eyre H, Mallam S, Torosyan N, Erickson KI, Lopez OL, Becker JT, Carmichael OT, Gach HM, Thompson PM, Longstreth WT & Kuller LH. (2016) Longitudinal Relationships between Caloric Expenditure and Gray Matter in the Cardiovascular Health Study. *J Alzheimers Dis*, 52: 719–729.

26. Raji CA, Meysami S, Hashemi S, Garg S, Akbari N, Ahmed G, Chodakiewitz YG, Nguyen TD, Niotis K, Merrill DA & Attariwali R. (2024) Exercise-Related Physical Activity Relates to Brain Volumes in 10,125 Individuals. *J Alzheimers Dis*, 97(2):829–839. https://doi.org/10.3233/JAD-230740. Online ahead of print.

27. Gutiérrez-Pérez IA, Delgado-Floody PD, Molina-Gutiérrez N, Campos-Jara C, Parra-Rojas I, Contreras-Osorio FH, Falfán-Valencia R, et al. (2024) Changes in Lifestyle and Physical and Mental Health Related to Long-Confinement due COVID-19: A Study during the First and Second Pandemic Waves in Mexico and Chile. *Psychol Health Med*, 29(1): 174–190. https://doi.org/10.1080/13548506.2023.2281295

28. Gottesman RF, Lutsey PL, Benveniste H, Brown DL, Full KM, Lee J-M, et al. (2024). Impact of Sleep Disorders and Disturbed Sleep on Brain Health: A Scientific Statement from the American Heart Association. *Stroke*, 55: e00–e00. https://doi.org/10.1161/STR.0000000000000453

29. Houldsworth A. (2024). Role of Oxidative Stress in Neurodegenerative Disorders: A Review of Reactive Oxygen Species and Prevention by Antioxidants. *Brain Commun*, 6(1):fcad356. https://doi.org/10.1093/braincomms/fcad356

4

Love for the Health of It

The healing power of touch is captured in the story behind the photo of the Rescuing Hug as is illustrated by Jaime Dubè in Figure 4.1. Twin sisters born

Figure 4.1
Illustration by Jaime Dubè

DOI: 10.4324/9781003462354-4

several weeks prematurely required neonatal intensive care unit (NICU) services. They were in separate incubators. As the stronger sister was thriving and the weaker sister was at risk for death, a nurse placed them in the same incubator. The stronger sister put her arm around the shoulders of her weaker sister who then began to thrive, and she survived. This case study of the outcome of an arm around a shoulder saving a dying twin sister's life is a powerful example of the potential healing power of human touch at an age before LOVE for the HEALTH of It was in these infants' vocabulary. A common practice now is to encourage "Kangaroo care" in which parents hold their newborn infant skin-to-skin against their chest for promoting health.

Each of us has memories of the healing power of another person's arm around our shoulders. One of the healing components on the farm of my youth was sleeping sisters snuggling together in double beds.

Profound positive purchases of love, "TLC," and touch were found in the research of Marian Diamond, PhD, the mother of neuroplasticity as described in her books *Enriching Heredity*[1] and *Magic Trees of the Mind*.[2,3] LOVE for the HEALTH of it included significant gains.

Love

When Marian Diamond spoke for the Washington State Psychological Association (WSPA) some years ago, we had the great fortune of hearing about her pioneering work. There's much more data in the Neurobics chapter on the ways the enriched environment she provided for her animals made significant improvements in the structure and function of their brain cells. At the WSPA meeting she said that when she told a conference of neuroscientists that she had evidence of "Enriching Heredity" even in the "elderly, 'a neuroscientist told her: "Young Lady, you cannot say your rats were elderly; ours live to be much older than yours."

PLEASE NOTE: Marian Diamond returned to her lab and made *only one change*. She instructed the lab technicians to remove the rats from their cage each day, hold them, and talk with them as is shown in Figure 4.1. Dr. Diamond found that *with this one change*, which she labeled *"TLC,"* her animals:

- Increased their lifespan by 50%;
- Lived to the equivalent of 90 human years; and
- Continued *"Enriching Heredity"* throughout their vigorous longevity!

Across time she referred to this saying: "We added LOVE."

Figure 4.2
As seen in the Davidson Films, Inc. documentary, "Older Brains, New Connections: A Conversation with Marian Diamond at 73," when Marian Diamond added "TLC," her animals enjoyed HUGE benefits in their life and brain health.

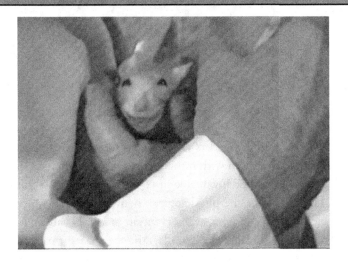

Touch and Talk

As many studies discussed herein indicate, Dr. Diamond was prescient when she wrote that she believed humans could also enjoy *"Enriching Heredity"* across a vigorous longevity.[3] This has been found in several subsequent studies. Her impressive findings with the single addition of "TLC" surely also qualifies her as the mother of Positive Psychology[4] which has shown that positive emotions promote health.[5–7]

This empowers us!! "TLC" of holding, touching, and talking are gifts we can easily share almost anywhere. As seen in Figure 4.2, these are some special people in my life sharing the TLC of touch, talk, and an arm around shoulders.

This was pre-COVID visit sharing love and the comforting arm around the shoulder!! Hereafter let that include remembering the potential for huge neuroplastic purchase which might continue to be among our benefits from touch and talk for nine decades.

Theme and variation of how we can bring this into our communities is heart-warmingly portrayed in multiple vignettes by Izmaylov[8] in her article: "The

Figure 4.3
Cousins could cuddle pre-COVID to share the love of touch and talk that empowers us and builds better brains.

Ways We Whisper Love." Supportive listening can be another way to enhance cognitive resilience even with elders who have decreased brain volume.[9]

Artificial intelligence used with human creativity, imagination, judgment, and problem solving will be invaluable in collecting data on the influence of LOVE for HEALTH for the physiological, neurocognitive, emotional, and behavioral factors of this important intervention in NICU care and across the lifespan. Artificial intelligence can handle variance analysis; pattern recognition; image analysis; automation; and information processing; humans can provide reasoning, judgment, imagination, creativity, and problem solving.[10] Thus, human efforts using artificial intelligence could accelerate our personalized "TLC" for *Enriching Heredity* by driving neuroplasticity in a positive direction and artificial intelligence could help evaluate the extent to which this LOVE for the HEALTH of it might influence the Flynn effect of finding the 20th century was "dominated by massive IQ gains from one generation to another" as documented in "at least 34 nations."[11] As refinements evolve, artificial intelligence can assist in bringing these benefits to *more* than 34 nations.

Adding this form of TLC is a touching way (pun intended) for us to begin in the vein of non-pharm. Liliana Alberto, MD, is artfully including this concept in a brilliant and sustainable way in Argentina as she encourages pregnant women and new mothers to breastfeed their infants. That's a very special way to touch and talk. A study of 571 adolescents "found that duration of exclusive breastfeeding was associated with cortical thickness in the superior and inferior parietal lobules."[12] In these adolescents, the longer duration of breastfeeding was also associated with higher "full scale and performance IQ." That adds to our appreciation of the special impact of touch and talk for the HEALTH of our neurons.

Listen

Remain open to listen when another person is ready to share. Anyone who knew when and how to be silent while listening versus when and how to elicit giggles has stayed on my list. If they also knew how to express care and concern without mentioning the stressor is an added gift of *serious* value!

> In a World Class show of compassionate support, she walked quietly behind me with a brief touch of her hand on my shoulder, smiled as I looked up to see the source of this TLC, had a barely perceptible pause in her stride as we maintained eye contact long enough for me to indicate if I needed time to talk, and she then turned as she walked on by.
>
> Susan Johnson, BSN/RN, expertly showing me compassion.

Over time, I had learned to train friends in what Susan did naturally and in other skills. I have offered them multiple-choice options between the two extremes modeled by Joan Rivers and former New York City mayor Ed Koch.

When asked something specific and painful, Rivers was noted for putting the back of her hand to her forehead as she semi-moaned: "Don't ask!" Her style of saying that was comedic; but that response has its value. The politician Koch was famous for greeting constituents with the generic question: "How'm I doin'?"

Those are invaluable models of how to be a good friend during tragedy and especially if the stress is long lasting and unavoidable. For starters, do not ask specific questions about that stressor. And smile while making ample use of your unique version of the generic question of Koch, such as "How's your world?"

That simultaneously lets your friends know you care and are open to wherever your friends feel a need to reside within their experience of their world at that moment. At the same time, it does not create an unpredictable drip of the painful topic that can bring tears to the eyes of your friend who is trying to remain happy and functional on the job and in the community.

Most folks have welcomed these options. Those who refused to avoid direct questions on tragic topics have never heard another word from me on any painful issues. Several said: "But I need to ask because I care!" These deeply concerned, compassionate, service-oriented friends have been grateful to realize that talking selectively and talking less about tragedy and stress affords all concerned with better brain health and more brain power while on task AND that this could enhance the immune response, the general health, and the healthy enhancing of neuroplasticity in the brains of all concerned.

I KNOW THAT YOU BELIEVE
YOU UNDERSTAND WHAT
YOU THINK I SAID,
BUT
I'M NOT SURE YOU REALIZE
THAT WHAT YOU HEARD IS
NOT WHAT I MEANT.

Author Unknown

Volunteer

The ways that an individual can use their love of others to participate in the increased use of Positive Psychology might be limited only by their imagination and creativity. Examples might include being a volunteer because staying socially active stimulates your brain to stay fit as was found in the study that found improved neurocognitive functioning in people who volunteered their services in the Baltimore Schools.[13] While in the schools to assist students needing support, their volunteer services enhanced their own neurocognition.

Ryan Hreljac[14] is an excellent example of creative volunteer efforts on behalf of others. As reported at www.RyansWell.ca, he was six years old when he had the vision, mission, drive, and other personal qualities that fueled this remarkable young man's drive to help others. When he learned that some African people without access to clean water were suffering many illnesses, he resolved to raise $70 to build a well for one small village to relieve their suffering. In four months, he met his goal. He is now Founder and Executive Director of Ryan's Well Foundation. Sharing this information and videos with youth could encourage and empower them.

Find a service organization that resonates with how you want to volunteer your time and talents. For example, Rotary International is a network of 1.4 million people focused on such things as ending polio and malaria. Their moto is service above self. MusicMendsMinds.org is an intergenerational effort to use music for improving neurocognition, mental health, and wellness. They believe that music is medicine.

PlayingForChange.org unites musicians from around the world in creating music to promote peace; on their website their video *The Weight,* is a heartwarming example of how they create and use music in service to others; the words in this song include "Take the weight off Fanny. Put the weight right on me."

Despite the power of love, adversity happens. For some, it is horrific.

Adversity

Out of the night that covers me,
Black as the pit from pole to pole,
I thank whatever gods may be
For my unconquerable soul.
In the fell clutch of circumstance
I have not winced nor cried aloud.
Under the bludgeonings of chance
My head is bloody, but unbowed.
Beyond this place of wrath and tears
Looms but the Horror of the shade,
And yet the menace of the years
Finds and shall find me unafraid.
It matters not how strait the gate,
How charged with punishments the scroll,
I am the master of my fate,
I am the captain of my soul.

William Ernest Henley

It is regrettable that so many of our global citizens are subject to adversity that is horrific. With the words of Henley above we read of horrific while he also provides the reader with some essential adjustments that he modeled in writing: "I am the master of my fate, I am the captain of my soul." While being grateful if we are dealing with less adversity, let us consider some of the options needed in dealing with difficult situations.

Self-Empowerment

This critical focus on self-empowerment is one key to using adversity to learn new skills. Let us never abandon addressing human problems. In solving and/or repairing these problems let us do our very best; then let's learn from the rest. I continue to teach that each of us needs several people with whom we can share our agony, maybe cry for a few minutes before easing away from the intensity of the pain and toward a happier stance. A brilliant professional colleague said: "Tragedy shared is divided. Joy shared is multiplied." Empowerment to express agony, wisdom in choosing with whom those intense emotions are laid bare, and TLC of self and others are all part of self-empowerment in the service of protecting and nourishing our precious neurons.

An important confounding variable to consider is human perception. The Baltimore Longitudinal Study of Aging assessed stereotypes in people without dementia decades prior to when annual magnetic imaging and brain autopsies were done.[15] Their findings were rather motivating (or sobering, depending on your perspective): "Those holding more negative age stereotypes earlier in life had significantly steeper hippocampal volume loss and significantly greater accumulation of neurofibrillary tangles and amyloid plaques, adjusting for relevant covariates." The cognitive skills of verbal fluency and memory[16] were also found to decline over two years associated with "negative perceptions of aging."

Thus, self-empowerment could include a positive age stereotype across your lifespan. The impact of stress that would be associated with negative perceptions was not addressed in these studies. However, these scientific reports strengthen the value of Diamond's report on neuroplastic purchases of "TLC" and suggest that TLC might need to begin with love of, and belief in, self-empowerment.

Forgive

One more example of the horrific most of us will never know. Also, an amazing response.

"When I received the call from Mike, my brother, on New Year's morning, I was stunned. 'Something terrible has happened,' he said. 'Mama's been murdered. There was blood on the carpets, the walls ...'" Everett Worthington wrote these words in his book,[17] *Five Steps to Forgiveness*. Most of us will avoid such horrific news but all can benefit from his wisdom within his book outlining his REACH model for reducing an agony that one could choose to forgive.

- **R**ecall the hurt ... to transform the experience into gold.
- **E**mpathize ... to achieve compassion for the transgressor.
- **A**chieve altruism ... to give the gift of forgiveness for the benefit of the other.
- **C**ommit publicly to forgive ... to release any doubt in your mind that you have forgiven.
- **H**old onto forgiveness ... by rehearsing to the point of mastery.

WOW! Henley and Worthington[17] provide examples of overcoming the horrific in the interest of returning to health. Both are inspiring models of self-empowerment and forgiveness that we can learn from.

Reframe

Albert Ellis is another source of inspiration. As the creator of rational emotive behavior therapy, an early form of cognitive behavior therapy, he wrote that he *"really* became a therapist"* to work on his own performance and social anxiety.[18] During a series of long and severe (*much less so* than described above) traumatic experiences, only two alternatives occurred to me in reaction to the teachings of Ellis: Either his perspective was a crock of stuff, or I was a miserable failure at being rational. Not being comfortable with either choice, I gave him a thumbnail sketch with no identifying data so he could not guess I was talking about myself. His response is instructive: "There are some outcomes in life that are truly tragic. They warrant strong emotions." Cognitive behavioral therapy was never meant to deny real pain resulting from real tragedy. It is meant to decrease the intensity, duration, and frequency of these emotions to help until the storm passes. Ellis's kindness and compassion in that evanescent moment has often helped me turn lemons into lemonade; it has been equally helpful in helping me help others in need of such assurances. Many studies have shown that cognitive behavioral therapy is at least as effective as prescribed antidepressants.[19]

As a personal example of a positive reframe, I started telling supportive friends that I had so much stress that I feared having a nervous breakdown if I didn't see a psychologist, so I looked in the mirror, but it didn't help. After several shared moments of laughter at various times with one friend or another, I reframed that to: "so I looked in the mirror, made direct eye contact with myself, and said, 'Thank you for helping me help me.'" Because we often know best how to learn from the best of it and harvest the rest of it. Usually, we know how to move from the cause of the stress to the elements that are empowering and to reframe the situation in ways that provide reason(s) for gratitude.

Often clarity of communication rests more in the mind of the speaker than in the awareness of those with whom they wish to share their thoughts. One of our experts on influences on our perceptions is Ellen Langer. She wrote a paper "Believing Is Seeing"[20] showing that "mind-manipulation can counteract physiological limits imposed on vision." Being accustomed to the traditional chart that had large letters at the top with letters becoming progressively smaller, individuals saw more when their eyesight was measured with the reversed order, i. e., smallest letters at the top and getting progressively larger. Manipulation of their mind-set counteracted "physiological limits imposed on vision."

Concern about human problems is essential and understandable. Negative emotions can interfere with cognition. For example, 40 healthy college students were assessed via EEG and event-related potential (ERP) during a recent study.[21] ERP "can reflect brain activity at a millisecond level" to assess cognitive processing and emotion. After 2–4 seconds of a neutral screen, they had intervals of negative and positive stimuli. Their cognitive performance was impaired, and their reaction time was significantly slower when the stimuli elicited extremely negative emotions.

Social Support

Social support can be emotional. It can be functional. It is a positive influence on neuroplasticity across the lifespan.

Diffusion MRI studies were collected on 103 newborn infants aged 2–6 weeks old during natural sleep.[22] Mothers completed online questionnaires when these infants were 6, 12, and 18 months old. Their results "offer evidence that various white matter tracts involved in emotion show early differentiation, specifically in terms of fear." They noted that "infants who showed both higher initial levels of fear and an increase in fear at a steeper rate had lower white matter microstructure." Anger and sadness did not show this pattern. This is an interesting beginning for researching the neurodevelopmental aspects of emotion.

Members of the IMAGEN Consortium group Analyzed data from 957 participants at ages 14 and again at 19 years old with the focus on "peer problems, emotional symptoms, and regional brain volumes."[23] Emotional symptoms and peer problems across adolescence change together. Higher levels of anxiety or mood disorders changed the grey matter volume in the ventromedial prefrontal cortex with greater impact than peer problems. Evidence-based interventions to relieve strong emotional responses to perceived negative life events could help protect this grey matter volume. Family support could be a factor. Since participants in IMAGEN were "recruited from multiple European countries" this could increase how these findings are generalized. Another study found that brain age was significantly increased along with serious losses in cognitive functioning in midlife in individuals with depression who were periodically assessed across 25 years.[24]

Diffusion weighted imaging and resting state fMRI were collected on 246 individuals.[25] They found "evidence for dampening of amygdala-prefrontal emotional control regions through adolescence, particularly in CM (centro-medial) amygdala and rACC (rostral anterior cingulate), paralleling known decreases in the influence of affect on behavior into adulthood." Since greater depression and anxiety in youth influenced this decreased influence of emotions on adult behaviors, they suggest that these factors can influence evidence-based intervention choices and efficacy during adolescence.

A person's perception of stressful events and of their capacity to cope with them influences neuroplasticity including "intra-cortical myelin (ICM), which is an important microstructure of brain and is essential for healthy brain functions."[26] MRI data on 1076 healthy adults aged 22–35 years and behavioral measures of "perceived chronic stress" and social support were assessed. They found that "both emotional support and instrumental support were negatively associated with perceived chronic stress, suggesting the protective effect of social support on stress."

After the Kumamoto earthquake when adults 65 years and above were relocated, their type and amount of physical activity was influenced by social support.[27] Being 75 years old and older and not participating in community activities were significantly associated with less physical activity. The researchers recommended active participation in community activities as well as receiving and giving social support in healthy activities. This is especially important when an individual moves to a new community given the evidence base on the value of social support.

A cross-disciplinary group considered the current data on evidence-based interventions for strengthening emotional well-being.[28] Summarized insights from convening "were: (1) existing interventions should continue to be adapted to achieve a large-enough effect to result in downstream improvements in psychological functioning and health, (2) research should determine the durability of interventions needed to drive population-level and lasting changes, (3) a shift from individual-level care and treatment to a public-health model of population-level prevention is needed and will require new infrastructure that can deliver interventions at scale, (4) interventions should be accessible and effective in racially, ethnically, and geographically diverse samples." Since mindfulness-based stress meditation and stress reduction can influence epigenetics as well as neuroplasticity, they can influence improved physical health, possibly with "intergenerational transmission of well-being, which could be viewed as a particularly durable intervention effect."

Art and Science of Human Flourishing was a novel elective college course.[29] They defined flourishing as leading an engaged, fulfilling, meaningful life that benefits others as well as benefiting self. After completing the course, 217 college students "reported significantly improved mental health (i.e., reduced depression) and flourishing, improvements on multiple attention and social-emotional skills (e.g., attention function, self-compassion), and increases in prosocial attitudes (empathic concern, shared humanity" compared to the control group of students not taking the course.

Gratitude

Gratitude comes easily. One of the gifts from my many geographic relocations is having many friends on several continents. Although many of them have encouraged me to call 24/7 as needed, it is often sufficiently healing to sit in solitude and review how a specific person(s) would express their unique perspective of social support.

A systematic review of 64 studies found a variety of gratitude interventions including writing or verbally expressing gratitude to others; thinking of things to feel grateful about; publishing photos with words of gratitude; visits; training; gratitude diaries; and conversation programs.[7] Control groups had other activities. Despite the risk of bias and other limitations, results encourage gratitude interventions. Participants in these interventions had less anxiety, less depression, more gratitude, better mental health, and more satisfaction with life. Their other benefits were "more positive moods and emotions, greater appreciation and optimism, more prosocial behavior, less worry, and less psychological pain." Gratitude letters increased happiness that lasted for one month. Participants had one week to write about their gratitude and then reading their gratitude letter to someone who had never been thanked for being especially kind to them; they read it with full expression of their own emotions; they then left this letter with this person they were thanking. Since this activity was associated with greater happiness that persisted for a month, that might be a good reason to write a gratitude letter each month.

Various methods of expressing gratitude were evaluated with 958 Australian adults.[30] Written letters and essays on any gratitude topic of their choice were associated with more subjective well-being than after writing gratitude lists. When gratitude letters were written to a specific person in their life, these writers had even greater feelings of other positive emotions in addition to more gratitude.

Online assessments and instructions were collected on 671 Australians aged 18–84 who were randomly assigned to an intervention condition.[31] All four conditions were "socially desirable activities (i.e., prosocial behavior, kind acts for oneself, extraverted behavior, open-minded behavior)" which they engaged in across 15 days and had "measured various emotions using momentary assessments." When doing prosocial behaviors, they reported greater self-confidence, greater sense of competence, and more meaning suggesting significant personal gains in community service. Doing kind things for others also increased their sense of connection. This report of "unique benefits of prosociality" is of particular interest coupled with the findings of enhanced neuroplasticity in trainers who participated in the Baltimore Experience Corp study that found increased cortical and hippocampal brain volumes in the trainers after they volunteered their services in the Baltimore schools.[13]

Prosociality

The National Institutes of Health (NIH) in the USA funded research networks in 2021 to study emotional well-being in families, clinically, and how this is reflected in brain aging.[32] Emotional well-being looked at networks as

> a multi-dimensional composite that encompasses how positive an individual feels generally and about life overall. It has both experiential features such as the emotional quality of momentary and everyday experiences and reflective features such as judgments about life satisfaction, sense of meaning, and ability to pursue goals that can include and extend beyond the self. These features occur in the context of culture, life circumstances, resources, and life course.

Typical elders were noted to avoid negativity, increase positive perspectives, and be less emotionally aroused. Strength versus vulnerability was a functional factor. Stress, locus of control, life satisfaction, and having a sense of purpose were other factors.

Let us welcome the paradigm shift in psychology from defining pathology and repairing liabilities and weaknesses. It is good for brain health that we follow this paradigm shift of Positive Psychology[4] to focusing on building strengths, improving self-esteem, being empathic, celebrating friendships, expressing love, setting goals, making commitments, fostering creativity, achieving, being mindful, benefiting from our spiritual beliefs and connections, having a sense of humor, increasing resiliency, enhancing optimism, solving what can be solved of stress and harvesting what we can from the rest, learning as we transition stress into "another opportunity to grow new skills," and significantly increasing well-being.

Positive Psychology researches these factors, techniques, and concepts to help us achieve the most fulfilling and flourishing life we can imagine which includes celebrating and using our optimal medical health, positively evolving optimal brain power across our entire lifespan, creativity, emotional intelligence, compassion, hope, talents, abilities, wisdom, altruism, volunteerism, forgiveness, gratitude, strengths of character, moral motivation, intrinsic motivation and flow, social support, meaning and purpose in life, prosocial behaviors, assets, humility, and any other strengths that have the potential to enhance neuroplasticity for a vigorous longevity for all. Adding the benefit of touch and talk as reported in the research of Marian Diamond highlights the wisdom of this paradigm shift in psychology to focus on behavioral impact on enhancing neuroplasticity. While no measure in her animals could estimate the neuroplastic purchase of this next way to use Positive Psychology, it's value in humans in huge.

> Hopelessness and despair threaten health and longevity. We urgently need strategies to counteract these effects and improve population health. Prosociality contributes to better mental and physical health for individuals, and for the communities in which they live. We propose that prosociality should be a public health priority.[33]

Since prosociality improves individual physical and mental health, these researchers' proposal to make prosociality a priority in public health is well-founded and urgent. Artificial intelligence working with a multidisciplinary team of researchers and clinicians can refine research protocols and prosociality interventions that assess and serve our diverse populations to maximize neuroplasticity and well-being for the benefit of our global population, of our academic institutions, or our healthcare, and of our corporate communities.

Laughter

Laughter is another form of promoting many health factors.[34] Their review considered eight studies with 315 participants whose average age was 38.6 years. Laughter interventions were associated with significantly reduced cortisol levels even when only one session of laughter was used. Another example of intervening during a horrific situation was described by Petrov & Marchalik when on "December, 2022, a viral video of a female Ukrainian soldier dancing on a snowy bank made headlines.[35] The cheerful choreography and the upbeat Ukrainian song were a stark contrast to the staccato of gunfire in the background." Some military personnel strongly supported this as they "recognized a reality of military life and trauma, and the need for finding an emotional outlet." A very small study with nursing students during the pandemic found reduced depression with online sessions of laughter therapy.[36] Even simulated laughter could have health benefits. Spontaneous or planned laughter sessions can be added to your list of healthy behaviors and can be done in groups. Tita Begashaw led a Tee Hee Laughter Group at Harborview Medical Center open to all and potentially good for all. During the COVID-19 lockdown, she still elicits giggles from behind her mask and from ten feet away for all who pause to share in her love and laughter (Figure 4.3).

"Doc, I see you got a haircut."
"Nope!
Got 'em all cut."

James Gottesman, MD, Urologist extraordinaire

And Master of Comedic Relief

Yoga

The Intelligent Internet of Medical Things (IIoMT) assessed the impact of Mano Shakti Yoga (MSY) on anxiety in university students.[37] The experimental group had an hour of Mano Shakti Yoga twice a week, while the control group continued to do their usual activities. Data from the smart belt was analyzed with machine learning algorithms. There was a significant reduction in anxiety in the experimental group compared to controls. This small study provides another example of how artificial intelligence can increase our database on effective interventions.

In a small study with healthy people whose average age was 27 years, their cognitive-motor interference was studies.[38] This refers to the extent to which cognitive resources are diminished for either motor or cognitive tasks when an activity requires both cognition and motor activity. Cognition was tested with serial subtraction which requires mental math and memory. Individuals who reported that they had practiced yoga two or three times a week for up to an hour for more than a year were compared to individuals of similar demographics who did not practice yoga. They were then tested on their correct responses in the serial subtraction during three variations of balancing tasks.

These researchers opined that "Yoga practitioners expended fewer motor and cognitive resources, thus, significantly reducing CMI during dual-tasking than non-practitioners, suggesting long-term Yoga practice could effectively promote the ability to maintain optimal function under challenging dual-tasking conditions." Although this is a small study, it focuses on an area that warrants further research since falling has the potential to cause brain injury with associated losses in neurocognitive functions.

A review of 89 studies found that yoga enhances neuroplasticity, protects brain cells from aging in response to stress and depression, reduces serum cortisol, and increases BDNF in healthy people and in depressed patients.[39] They opined that yoga and meditation "can prevent initiation of drugs/medications to tackle stress, anxiety, and depression in vulnerable populations."

Mindfulness Meditation

Since some training in meditation might contribute to healthy aging through influence on telomeres, "meditation training may impact acute and habitual stress responses as pathways to improved cell aging."[40] Telomeres are the tiny protective ends of each of your DNA strands. Since telomere length is considered a form of measuring biological aging as well as being related to severe stress, researchers looked at the relationship between telomere length and their participants' pattern of staying present as compared to letting their mind wander. Shorter telomeres were found in women who experienced a lot of mind wandering, than was true of women who reported low mind wandering. This was the case despite the level of stress. These researchers suggest that a healthier biochemistry and better cell longevity may result from maintaining attention on the present. Being mindful and focusing on the present moment are components of meditation that have been associated with improvements in mental, emotional, and physical health.

Perhaps the world expert in using thought alone to improve brain plasticity is Richard Davidson whose study of Buddhist meditation practitioners with 15–40 years of meditation practice "found robust gamma-band oscillation and long-distance phase-synchrony during the generation of the nonreferential compassion meditative state."[41] After eight weeks of mindfulness meditation, participants had increased brain activation that "predicted the magnitude of antibody titer rise to" a vaccine.[42] Davidson reviewed research findings that "underscore the structural plasticity of emotional circuitry in response to both acute and chronic stress, particularly alterations of spine density and dendritic length and branching in hippocampus, amygdala and prefrontal cortex."[43,44] He studied gurus of meditation who had greater brain volume in their prefrontal cortex, the part of your brain involved in processing complex thoughts and decisions. Since inflammation has been associated with many conditions that include mental decline, it is important to learn that mindfulness meditation could reduce inflammation like prescription drugs do. A faster response was seen in experienced meditators. After eight hours of intense meditation, in experienced meditators they "identified 61 differentially methylated sites (DMS) after the intervention. These DMS were enriched in genes mostly associated with immune cell metabolism and ageing and in binding sites for several transcription factors involved in immune response and inflammation."[45] The control group did not show any of these changes. These researchers opined that "a short meditation intervention in trained subjects may rapidly influence the epigenome at sites of potential relevance for immune function and provide a better understanding of the dynamics of the human methylome over short time windows."

More information on the research of Richard Davidson and colleagues is available at CenterHealthyMinds.org where his many publications researching the neural bases of emotions are available. In a video there he reports that His Holiness the Dalai Lama challenged the perspective of research on negative aspects of the human condition and suggested studying the healthy components of our human conditions.

In the 1980s when the Dalai Lama asked what relationship "could there be between Buddhism, an ancient Indian philosophical and spiritual tradition, and modern Science?" this "led to the creation of the Mind & Life Institute" which studies contemplative science. He then launched "contemplative neuroscience" in which scientists study the brain activities of Buddhist meditation experts who had practiced meditation for more than 10,000 hours. Their research shed light on "why this set of techniques for training the mind holds great potential for supplying cognitive and emotional benefits" which can include greater wellness and feeling calmer. They also found: "Physiological changes in the brain – an altered volume of tissue in some areas – occur through meditation. Practitioners also experience beneficial psychological effects: they react faster to stimuli and are less prone to various forms of stress." Much of the research reviewed was on three forms of meditation: compassion and loving-kindness meditation; mindfulness meditation; and focused attention meditation.

People more than 60 years old were involved in a study using "mindfulness-based cognitive therapy (MBCT) for eight weeks.[46] The control group had treatment as usual. Their symptoms were evaluated with the Hamilton Anxiety Scale (HAMA) and the Hamilton Depression Scale (HAMD). Resting-state and structural and functional connectivity were measured with MRI. The relationship between clinical symptom improvements and brain connectivity changes was evaluated. Individuals receiving MBCT had significantly lower scores on their HAMD after the eight weeks and at the three month follow up compared to the control group. "Diffusion tensor imaging analyses showed that fractional anistrophy of the MFG-amygdala significantly increased." These researchers opined "That MBCT improves depression and anxiety symptoms" associated with late life depression, that MBCT strengthened functional and structural connections between the amygdala and the MFG, and this increase in communication correlated with improvements in clinical symptoms."

Researchers compared long-term meditators to controls between the ages of 24 and 77 years. Practitioners of long-term meditation may have less age-related loss of brain gray matter (where the brain cells are). So, meditation could protect your brain by reducing age-related brain cell decline. This could benefit the structure throughout your brain. Even improvements in white brain matter have been found with short periods of meditation. White brain matter is the axons, insulated branches off the cell body that connect to other cells. Research on compassion meditation training found that greater compassion for others might include the benefit of decreased depression. Thus, it could be mutually beneficial for mood and brain matter, for yourself, and for those whom you focus on with compassion.

Another benefit of my living and traveling overseas is the experience of meditation, yoga, and the perspective of individuals who used these experiences on a regular basis. They expressed and demonstrated the benefits in ways that simply reading doesn't. However, the findings in research are compelling.

Healthy male meditators with an average age of 43 who had meditated daily for about seven years had longer telomeres compared to a similar group that had not meditated.[47] Telomeres are protein structures at both ends of a chromosome to protect and preserve our genetic information. They opined that the finding of longer telomeres in the individuals who practiced daily meditation suggests that this lifestyle practice of frequent meditation has the potential to promote healthier aging with the multilevel benefits of meditation noted in telomere dynamics.

Forty-seven children between the ages of nine and eleven who had problems with attention and/or anxiety were assessed by MRI and tests for anxiety prior to and after they were given ten weeks of mindfulness training.[48] One of the tests was the CPT 3 which "measured four aspects of attention including sustained attention, inattentiveness, impulsivity, and vigilance, via 360 trials presented in 14 minutes. Variables measured included detectability, errors of commissions, omissions, and hit response time." Mindfulness training included encouraging them to let their thoughts come and go; they were also taught to focus their attention on their breathing. "Mindfulness skills were developed through concepts that included paying attention, recognizing worry, letting go of thoughts, dealing with difficult emotions, and gratitude." Their different ethnicities included "Caucasian 94%, Asian 4%, Aboriginal and Torres Strait Islander 2%." After the mindfulness training, they performed better. They also had greater functional connectivity in the Salience Network and in the Frontoparietal network which "were further associated with improved behavioral outcomes post-intervention." These researchers opined that their "findings suggest mindfulness strategies may be an effective intervention for sub-clinical anxiety-related attention impairments in pre-adolescent children, however, need to be considered in the context of developing functional connectivity."

> *The secret of health for both mind and body is not to mourn for the past, worry about the future, or anticipate troubles but to live in the present moment wisely and earnestly.*
>
> *Buddha*

Brain Reward System

Mobilizing the reward system of the brain might protect resilience to stress and improve health outcomes related to stress (Anderson et al., 2023). Activating the brain reward system reduces the stress response and is associated with better health including less depression.[49] Artificial intelligence with human guidance could increase knowledge about neuroanatomical and neurobiological factors in the reward system to refine strategic evidence-based interventions to maximize enhancing neuroplasticity.

> "We don't stop playing because we grow old;
> we grow old because we stop playing."
>
> *George Bernard Shaw*

Take some comfort in knowing that research supports this refreshing perspective! Focus on and celebrate your unique strengths rather than spending much of your time correcting what you label as your weaknesses. In this way you can make your strengths productive and your weaknesses irrelevant. Ask what the world needs, of course. There is a lot of health in putting service above self. Ask, also, what makes your heart sing and lifts your face in a smile. Do that. Because people like to do what they do well. For the rest of it, ask: "Who can do this?" Much of life takes a team; some things require a village; and the social unrest as well as the pandemic require evidence-based efforts of our entire global community.

Research has been identifying choices focused specifically on increasing happiness. By keeping a list of positive experiences and sharing them with close friends, happiness increased, especially when the listener responded with enthusiastic support.

COVID-19 has changed so much about how we maintain and support our friendships. For example, condo COVID visiting is more like smiling with comfort knowing "they" are home as you walk by the doors smiling at their shoes as seen in Figures 4.4 and 4.5.

> "Three things in human life are important. The first is to be kind.
> The second is to be kind.
> The third is to be kind."
>
> *Henry James*

Figure 4.4
The psychiatrist who saw patients in this office was told by one of his patients: "Doc, the last therapist I saw had an office so neat I felt inferior! I don't feel that way here."

Figure 4.5
COVID: couldn't cuddle. Could find comfort in knowing neighbors were home.

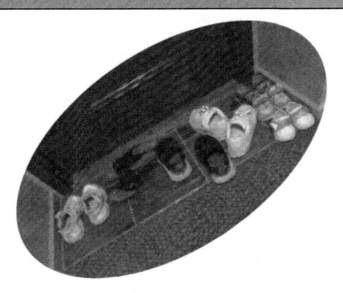

The ways that an individual can participate in the increased use of Positive Psychology might be limited only by their imagination and creativity. Examples might include being a volunteer because staying socially active stimulates your brain to stay fit as was found in the study that found improved neurocognitive functioning in people who volunteered their services in the Baltimore Schools.[13] Smile freely because it can transform another person's day in a positive way. Give compliments copiously because it helps you and others grow. Remain socially active because relations are pharmaceutical.

One of my earliest workshops was on loneliness. There was standing room only inside the room. The lonely crowd flowed out into the hallways. The large number of people suffering from loneliness is a public health concern because of the associated greater risk of health problems. There are several ways to thrive in love and protect your precious brain cells from the ravages of the chemistry of stress. Let your goals include using adversity to expand your skills in the many ways considered above in addition to any others that could make your heart sing.

Conditioned Relaxation Response

One of the best tools for doing this is progressive relaxation because this technique can increase your awareness of tension in your muscles before you might even be aware that your muscles are tense. You can create an audio version in your own voice that instructs you to tense and then release tension in opposing sets of muscles in your hands, arms, etc. Use it to develop a conditioned relaxation response. With a conditioned relaxation response, you have built the skill of taking yourself from misery to mellow in milliseconds even in a crowd of people as well as when faced with what initially seems unbearable. Building a conditioned response to a cue is a key element in turning stress into power. Most people can develop the conditioned response in about two weeks by using the progressive relaxation technique twice a day. Using it once a day just takes longer to become conditioned to respond to the cue with deep, deep relaxation. That conditioned response can help you build the skills to be calmer until the task at hand is resolved.

"A talent is formed in stillness; a character in the world's torrent."

Goethe

Aerobic Exercise (More on This in the Next Chapter)

In a related vein, reducing depression is a top priority. Duration of depression is more powerful than age in predicting loss of volume in the human hippocampus, the part of your brain that is associated with learning and memory. Aerobic exercise is at the top of the stress management, antidepressant, and mood elevating list of empowering skills.

"My Friend, you were right about aerobic exercise! At first, I had to force myself past grief and low energy to even get on the trainer. Then the saddle chafed so much I thought I just could not stay there no matter what! By force of will alone, I began to pedal the bike. But ... around ten minutes, it was like – I could see clearly now that the sun was gonna come up (Figure 4.6).
THEN ... around twenty-two minutes the sun did come up.
I felt so much better that I knew I could make it to the goal of thirty minutes.
AND THEN ...
as I was finishing the thirty minutes,

Figure 4.6
The greater the love the greater the grief. Fortunately,
aerobic activity can facilitate health and healing.

I remembered that you said staying to forty was sometimes even better.
So, I stayed. At forty minutes it was like the
SUN was GLORIOUS and the flowers were in bloom!"
A verbatim quote from a recent convert to the value of aerobic activity in
changing the chemistry of grief.
This gain in insight and empowerment occurred just days after learning of a
death in the family.

Discussion

There are centenarians with no dementia and good cognitive functioning in many countries. Incidence of dementia per calendar decade has decreased 13% in the USA and Europe in recent decades. The frequency of visits with their children was associated with maintaining cognitive efficiency in Chinese centenarians.[50] Since emotions are contagious, LOVE for the HEALTH of It might be the best, portable, brain-enhancing, free, gift you can share generously with your family, friends, and community.

> "Together, You Can Redeem the Soul of Our Nation.
> I urge you to answer the highest calling of your heart and
> stand up for what you truly believe.
> When historians pick up their pens to write the story of the 21st century, let
> them say that peace finally triumphed over violence, aggression, and war.
> So, I say to you, walk with wind, brothers and sisters, and let the spirit
> of peace and the power of everlasting love be your guide."

John Lewis

Written in the last essay in the final days before he died, 7/17/2020.

Love, TLC, peace, and happiness might be the most portable and easily shared non-pharm ways to increase brain power by turning adversity into stress power. Twenty-four-seven you can bring a smile to your face to share happiness that can be contagious in your community. That is behind my motivation described earlier encouraging you to enjoy playing at (rather than working at) the lifestyle changes you choose.

As the two examples at the beginning of this chapter illustrate, the power and many significant purchases of love, "TLC," talk, and touch are worth repeating for emphasis. Marian Diamond,[1,2] the mother of neuroplasticity and of Positive Psychology, *first* provided pivotal data on how an enriched environment (more on that in the chapter on neurobics) improves the structure and function of brain cells. *Then* with the solitary change of having her lab technicians take the animals out of the cage, hold them, and talk with them ... a change she labeled "TLC" ... the benefits included a 50% increase in lifespan to the equivalence of 90 human years AND continued *"Enriching Heredity"* for the duration of the equivalence of 90 human years.[3] Subsequent research shows that Dr. Diamond was prescient in saying that these same gains in human neuroplasticity would be possible across a vigorous and extensive lifespan. The neuroplastic purchases of love, "TLC," are significant. As the picture of the Rescuing Hug[51] shows, the power of the arm around the shoulder to save the life of the twin sister who was at risk for death is similarly empowering. Touching "TLC" must be part of ALL lives for hacking neuroplasticity for healthy brain aging!!

Hacking neuroplasticity with love, TLC, touch, and social support are essential lifestyle choices as these two examples make abundantly clear. That is why **LOVE** is at the top of our list. Although stress happens to everyone, for some people it is horrific. Horrific requires more.

Humans create the chemistry of fight or flight to cope with real threat to life or limb. That is a good thing when it enhances your run for your life across the savannah to avoid being a lion's lunch. Close relatives to an individual who has traumatic brain injury (TBI) are also severely impacted by this comprehensive traumatic experience. Services will be required for the family as well as for the person with the TBI. Volunteers trained in Helping Others through Purpose and Engagement (HOPE) lent truth to the perspective that you cannot give without receiving. These HOPE trained volunteers experienced significant improvements in their life satisfaction and in their self-perceived success in helping others.

William Ernest Henley provided an exemplary model of responding to the horrific via *self-empowerment* in writing "It matters not how strait the gate, how charged with punishments the scroll, I am the master of my fate. I am the captain of my soul." Everett Worthington responded to the horrific by writing the REACH model of how to *forgive* in his book *Five Steps to Forgiveness*.[17] Albert Ellis[18] *"really* became a therapist" to heal himself and taught methods to *reframe* thoughts behind our depression and anxiety. A gentleman in a third course of chemotherapy, when told "I see you're losing your hair again," replied: "No. I'm getting a new hair style." Many researchers provide us with data on the multiple ways that having good *social support* has a positive influence on overall health and neuroplasticity; this includes empathic *listening*. *Gratitude* expressed in writing, in person, and online has been associated with improved mental health. *Prosociality* has shown so much positive influence on physical and mental health that researchers say it "should be a public health priority."[33] *Laughter* has many health factors[34] including reducing cortisol; laughter therapy can also be administered online. *Yoga* benefits for improving mental health have been measured using artificial intelligence and machine learning algorithms.[37] *Meditation* can reduce stress and influence positive changes in cognition and in the immune response.[45] *Progressive Relaxation* used with sufficient frequency to create a conditioned response can afford relief in the early stages of building tension with stress. *Aerobic Exercise* is a healthy way to work off the chemistry of stress that is essential for immediate response to some forms of stress.

Most of the time you probably use good conflict-resolving and/or problem-solving skills and/or superior human relations skills to get through trauma. Then you get back to the task at hand. Even if that was good judgment, superior time management and excellent personal relationship skills, stress can leave your chemistry ready for a race across the savannah to avoid being someone's lunch.

That is one of many reasons that wise business owners have a cardio machine readily available and reward employees for using it. That could get in the health-promoting aerobics early enough to not interfere with sleep; it could also add to the corporate bottom line by saving dollars that would have been spent on medical issues from too little exercise. Having multiple-choice comedic videos to watch while on the brief exercise break could add laughter to the day for additional health benefits. Another small room for meditation during breaks could also add to health benefits and save corporations money on healthcare benefits. Keep a poster on the wall where employees can write their personal reasons for gratitude. Artificial intelligence with responsible human creativity can help collect data on benefits of these evidence-based interventions.

On a personal note, friends across the years have told me they have never known anybody like me whose life has been filled with so many extremes. I believe some of my readers may have similar experiences. The negative extremes these friends were observing across my lifespan ranged from poverty through caring for medically fragile family and facing near death as well as untimely, sometimes even avoidable, deaths. Extreme positives include having gone around the World twice, living on Borneo Island as a Peace Corps dependent, and being one of the O2K family that bicycled in 45 countries on 6 continents throughout the year 2000. Several friends who have been close to me through the struggles and tragedies have declared me to be the world expert in managing stress; they also agree with me that mine has been a gifted life.

Skip this if it seems unacceptable in any way that part of my stress management has always included how I interact with friends. Eons ago I started collecting brilliant friends who were spontaneously positive, sensitive, compassionate, comedic, and available for hugs. Anyone who knew when and how to be silent while listening versus when and how to elicit giggles has stayed on my list. If they also knew how to express care and concern without mentioning the stressor, that is an added gift of *serious* value during times when we need to avoid the constant dribbling reminders of the stressor while we need to stay clear, cogent, and convincing on our important work of the day!! The example given previously of the World Class show of compassionate support by Susan Johnson is a supreme role model. We all need several special souls with whom we can discuss agony so it doesn't stay on recycle in our minds; maybe cry a few minutes; with whom we can also move from the angst to happier emotions, hopefully laughter; and with whom it is brain food to celebrate our strengths and successes.

To return from the personal to the science, cortisol, a stress chemical that can damage cells, and C-reactive protein (CRP), a measure of inflammation, was measured before and after teaching senior citizens ways of changing negative thoughts to a positive perspective. This intervention reduced cortisol measures. *Four years later* the CRP in these seniors was improved. Since many health issues such as cardiovascular disease, hypertension, diabetes, and dementia are associated with inflammation and stress, that reduction in CRP can also include a significant improvement in quantity and quality of life. Those are major purchases!!

The potential for aerobic exercise and these positive perspectives to enhance your brain in these ways is empowering to say the least. A time of stress is prime time to do some type of aerobic activity to work off that chemistry, get the antidepressant and mood elevating chemicals that aerobics produce and probably improve your sleep; however, timing is a factor. The changes described in the quote above imply that earlier in the day would be preferable so you can enjoy those benefits during your waking hours. This will also give your internal body temperature a chance to cool off again in the interest of sleeping well.

The healing power of the Rescuing Hug,[51] Service above Self, and the TLC in the Marian Diamond[1-3] lab are among the models we can use as we share LOVE for the HEALTH of it and use adversity to increase our loving, healing skills. Let us work, play, love, listen, volunteer, empower, forgive, reframe, be socially active, express our gratitude, emphasize prosociality, laugh, use the Intelligent Internet of Things for yoga, meditate, develop a conditioned relaxation response, and do aerobic exercise for the healthy ways of using LOVE for the HEALTH of it. United in this effort we can make our World a better place for ALL. That is pertinent to your individual brain cells, muscles, skills, etc. Comforted by the picture of the Rescuing Hug,[51] we can appreciate the demonstration of our human capacity shortly after birth to use the healing power of compassion and an arm around our shoulders. This might begin with compassion for self which could have benefits for brain activity as well as reducing negative emotions.[52]

Smile freely because it can transform another person's day in a positive way. Give compliments copiously because it helps you and others grow. Remain socially active because relations are pharmaceutical.

References

1. Diamond MC. (1988). *Enriching Heredity: The Impact of the Environment on the Anatomy of the Brain.* New York: The Free Press.
2. Diamond MC & Hopson J. (1999). *Magic Trees of the Mind: How to Nurture Your Child's Intelligence, Creativity and Healthy Emotions from Birth through Adolescence.* New York: Plume.
3. Diamond MC, Johnson R, Protti A, et al. (1984). Plasticity in the 904-day-Old Male Rat Cerebral Cortex. *Exp Neurol*, 87: 309–317.
4. Peterson, C. (2008) What Is Positive Psychology, and What Is It Not? *Psychology Today*, May 16. https://www.psychologytoday.com/us/blog/the-good-life/200805/what-is-positive-psychology-and-what-is-it-not? (Accessed online July 2, 2023).
5. Seligman MEP, Steen TA, Park N & Peterson C. (2005) Positive Psychology Progress: Empirical Validation of Interventions. *Amer Psychol*, 60(5): 410–421.
6. Diener E, Oishi S & Tay L. (2018). Advances in Subjective Well-Being Research. *Nat Hum Behav*, 2(4): 253–260. https://doi.org/10.1038/s41562-018-0307-6.
7. Diniz G, Korkes L, Tristão LS, Pelegrini R, Lacerda B & Bernardo WM. (2023). The Effects of Gratitude Interventions: A Systematic Review and Meta-Analysis. *Einstein (Sao Paulo)*, 21: eRW0371. https://doi.org/10.31744/einstein_journal/2023RW0371.
8. Izmaylov MI. (2023). The Ways We Whisper Love. *The Amer J Med*, 136(1). https://doi.org/10.1016/j.amjmed.2022.06.024
9. Salinas J, O'Donnell BA, Kojis DJ, et al. (2021) Association of Social Support with Brain Volume and Cognition. *JAMA Netw Open*, 4(8): e2121122. https://doi.org/10.1001/jamanetworkopen.2021.21122
10. Lawry T. (2023). *Hacking Health Care: How AI and the Intelligence Revolution Will Reboot an Ailing System.* New York: Routledge.
11. Flynn JR. (2018). Reflections about Intelligence over 40 Years. *Intelligence*, 70: 73–83. https://doi.org/10.1016/j.intell.2018.06.007
12. Kafouri S, Kramer M, Leonard G, Perron M, Pike B, Richer L, et al (2013) Breastfeeding and Brain Structure in Adolescence. *Int J Epidemiol*, 42: 150–159.

13. Carlson MC, Kuo JH, Chuang Y-F, et al. (2015). Impact of the Baltimore Experience Corps Trial on Cortical and Hippocampal Volumes. *Alzheimer's Dementia*. https://doi.org/10.1016/j.jalz.2014.12.005
14. Hreljac R. www.RyansWell.ca (Accessed online November 12, 2023).
15. Levy, BR, Ferrucci L, Zonderman AB, Slade MD, Troncoso J & Resnick SM. (2016). A Culture–Brain Link: Negative Age Stereotypes Predict Alzheimer's Disease Biomarkers. *Psychol Aging*, 31(1): 82–88. https://doi.org/10.1037/pag0000062
16. Robertson DA, King-Kallimanis BL & Kenny RA. (2016). Negative Perceptions of Aging Predict Longitudinal Decline in Cognitive Function. *Psychol Aging*, 31(1): 71– 81. https://doi.org/10.1037/pag0000061
17. Worthington E. (2001) *Five Steps to Forgiveness, the Art and Science of Forgiving: Bridges to Wholeness and Hope.* New York: Crown Publishers.
18. Ellis A. (2005). Why I (Really) Became a Therapist. *J Clin Psychol*, 61(8): 945–948.
19. Antonuccio DL, Danton WG, DeNelsky GY, Greenberg R & Gordon JS. (1999). Raising Questions about Antidepressants. *Psychother Psychosom*, 68: 3–14.
20. Langer EJ, Djikic M, Pirson M, Madenci A & Donohue R. (2010). Believing Is Seeing: Using Mindlessness (Mindfully) to Improve Visual Acuity. *Psychol Sci*, 21(5): 661–666. https://doi.org/10.1177/0956797610366543
21. Yang K, Zeng Y, Tong L, Hu Y, Zhang R, Li Z & Yan B. (2023). Extremely Negative Emotion Interferes with Cognition: Evidence from ERPs and Time-Varying Brain Network. *J Neurosci Methods*, 396: 109922. https://doi.org/10.1016/j.jneumeth.2023.109922
22. Planalp EM, Dowe KN, Alexander AL, Goldsmith HH, Davidson RJ & Dean DC. (2023). White Matter Microstructure Predicts Individual Differences in Infant Fear (But not Anger and Sadness). *Dev Sci*, 26(3): e13340. https://doi.org/10.1111/desc.13340.
23. Stepanous J, Munford L, Qualter P, Nees F, Elliott R & IMAGEN Consortium. (2023). Longitudinal Associations between Peer and Family Relationships, Emotional Symptoms, and Regional Brain Volume across Adolescence. *J Youth Adolesc*, 52: 734–753. https://doi.org/10.1007/s10964-023-01740-7
24. Dintica CS, Habes M, Erus G, et al. (2023) Long-Term Depressive Symptoms and Midlife Brain Age. *J Affect Disord*, 320: 436–441.
25. Jalbrzikowski M, Larsen B, Hallquist MN, Foran W, Calabro F & Luna B. (2023) Development of White Matter Microstructure and Intrinsic Functional Connectivity between the Amygdala and Ventromedial Prefrontal Cortex: Associations with Anxiety and Depression. Biol Psychiatry, 82(7): 511–521. https://doi.org/10.1016/j.biopsych.2017.01.008
26. Guo Y, Wu H, Dong D, Zhou F, Li Z, Zhao L & Long Z. (2023). Stress and the Brain: Emotional Support Mediates the Association between Myelination in the Right Supramarginal Gyrus and Perceived Chronic Stress. *Neurobiol Stress*, 22: 100511. https://doi.org/10.1016/j.ynstr.2022.100511
27. Kanamori Y, Ide-Okochi A & Samiso T. (2023). Factors Related to Physical Activity among Older Adults Who Relocated to a New Community after the Kumamoto Earthquake: A Study from the Viewpoint of Social Capital. *Int J Environ Res Public Health*, 20: 3995. https://doi.org/10.3390/ijerph20053995

28. Kubzansky LD, Kim ES, Boehm JK, Davidson RJ, Hufman JC, Loucks EB, et al. (2023). Interventions to Modify Psychological Well-Being: Progress, Promises, and an Agenda for Future Research. Afect Sci, 4: 174–184. https://doi.org/10.1007/s42761-022-00167-w

29. Hirshberg MJ, Colaianne BL, Greenberg MT, Inkelas KK, Davidson RJ, Germano D, et al. (2023). Can the Academic and Experiential Study of Flourishing Improve Flourishing in College Students? A Multi-University Study. Mindful (N Y), 13(9): 2243–2256. https://doi.org/10.1007/s12671-022-01952-1

30. Regan A, Walsh LC & Lyubomirsky S. (2023). Are Some Ways of Expressing Gratitude More Beneficial Than Others? Results from a Randomized Controlled Experiment. *Affect Sci*, 4(1): 72–81. https://doi.org/10.1007/s42761-022-00160-3.

31. Regan A, Margolis S, Ozer DJ, Schwitzgebel E & Lyubomirsky S. (2023). What Is Unique About Kindness? Exploring the Proximal Experience of Prosocial Acts Relative to Other Positive Behaviors. *Affect Sci*, 4(1): 92–100. https://doi.org/10.1007/s42761-022-00143-4. eCollection 2023 Mar.

32. Lin FV, Zuo Y, Conwell Y & Wang KH. (2023). New Horizons in Emotional Well-Being and Brain Aging: Potential Lessons from Cross-Species Research. *Int J Geriatr Psychiatry*, e5936. https://doi.org/10.1002/gps.5936

33. Kubzansky LD, Epel ES & Davidson RJ. (2023). Prosociality Should Be a Public Health Priority. *Nat Hum Behav*, 7(12): 2228. https://doi.org/10.1038/s41562-023-01717-3. Online ahead of print.

34. Kramer CK & Leitao CB. (2023). Laughter as Medicine: A Systematic Review and Meta-Analysis of Interventional Studies Evaluating the Impact of Spontaneous Laughter on Cortisol Levels. PLoS One, 18(5): e0286260. https://doi.org/10.1371/journal.pone.0286260

35. Petrov D & Marchalik D. (2023) Laughter Is the Best Medicine. *Lancet*, 401(10391): 1844. https://doi.org/10.1016/S0140–6736(23)01074-7

36. Ozturk FO & Tekkas-Kerman K. (2022). The Effect of Online Laughter Therapy on Depression, Anxiety, Stress, and Loneliness among Nursing Students during the Covid-19 Pandemic. *Arch Psychiatr Nurs*, 41: 271–276. https://doi.org/10.1016/j.apnu.2022.09.006.

37. Pal R, Adhikari D, Heyat BB, Guragai B, Lipari V, Ballester JB, et al. (2022). A Novel Smart Belt for Anxiety Detection, Classification, and Reduction Using IIoMT on Students' Cardiac Signal and MSY. Bioeng (Basel), 9(12): 793. https://doi.org/0.3390/bioengineering9120793

38. Subramaniam S & Bhatt T (2017) Effect of Yoga Practice on Reducing Cognitive-Motor Interference for Improving Dynamic Balance Control in Healthy Adults. *Complement Ther Med*, 30: 30–35. https://doi.org/10.1016/j.ctim.2016.10.012

39. Varne SR & Balaji PA (2023) A Systematic Review on Molecular, Bio-Chemical, and Pathophysiological Mechanisms of Yoga, Pranayama and Meditation Causing Beneficial Effects in Various Health Disorders. *Indian J Integr Med*. https://www.researchgate.net/publication/375888948

40. Conklin QA, Crosswell AD, Saron CD & Epel ES. (2019) Meditation, Stress Processes, and Telomere Biology. *Curr Opin Psychol*, 28: 92–101. https://doi.org/10.1016/j.copsyc.2018.11.009

41. Lutz A, Greischar LL, Rawlings NB, Ricard M & Davidson RJ. (2004). Long-term Meditators Self-Induce High-Amplitude Gamma Synchrony during Mental Practice. *PNAS*, 101(46): 16369–16373. www.pnas.org-cgi; https://doi.org/10.1073.pnas.040740110

42. Davidson RJ, Kabat-Zinn J, Schumacher J, Rosenkranz M, Muller D, Santorelli SF, et al. (2003). Alterations in Brain and Immune Function Produced by Mindfulness Meditation. *Psychosom Med*, 65: 564–570. https://doi.org/10.1097/01.PSY.0000077505.67574.E3

43. Davidson RJ & Lutz A. (2008). Buddha's Brain: Neuroplasticity and Meditation. *IEEE Signal Process Mag*, 25(1): 176–174.

44. Davidson RJ & McEwen BS. (2012). Social Influences on Neuroplasticity: Stress and Interventions to Promote Well-Being. *Nat Neurosci*, 15(5). https://doi.org/10.1038/nn.3093

45. Chaix R, Fagny, M, Cosin-Tomás, M, Alvarez-López, M, Lemee, L, Regnault, B, Davidson, RJ, Lutz, A & Kaliman, P. (2020). Differential DNA Methylation in Experienced Meditators after an Intensive Day of Mindfulness-Based Practice: Implications for Immune-Related Pathways. *Brain Behav Immun*, 84: 36–44. https://doi.org/10.1016/j.bbi.2019.11.003

46. Li H, Yan W, Wang Q, Liu L, Lin X, Zhu X, Su S, Sun W, Sui M, Bao Y, Lu L, Deng J, & Sun X. (2022) Mindfulness-Based Cognitive Therapy Regulates Brain Connectivity in Patients with Late-Life Depression. *Front Psychiatry*, 13: 841461. https://doi.org/10.3389/fpsyt.2022.841461

47. Dasanayaka NN, Sirisena ND & Samaranayake N (2023) Associations of Meditation with Telomere Dynamics: A Case–Control Study in Healthy Adults. *Front Psychol*, 14: 1222863. https://doi.org/10.3389/fpsyg.2023.1222863

48. Kennedy M, Mohamed AZ, Schwenn P, Beaudequin D, Shan Z, Hermens DF & Lagopoulos J. (2022) The Effect of Mindfulness Training on Resting-State Networks in Preadolescent Children with sub-Clinical Anxiety Related Attention Impairments. *Brain Imaging Behav*, 16: 1902–1913. https://doi.org/10.1007/s11682-022-00673-2

49. Dutcher JM. (2023) Brain Reward Circuits Promote Stress Resilience and Health: Implications for Reward-Based Interventions. *Curr Dir Psychol Sci*, 32(1): 65–72. https://doi.org/10.1177/09637214221121770

50. Yin S, Yang Q, Xiong J, Li T, & Zhu X (2020) Social Support and the Incidence of Cognitive Impairment Among Older Adults in China: Findings from the Chinese Longitudinal Healthy Longevity Survey Study. *Front Psychiatry*, 11: 254. https://doi.org/10.3389/fpsyt.2020.00254

51. The Hug That Helped Change Medicine – YouTube.

52. Luo S, Che X & Li H. (2023). Concurrent TMS-EEG and EEG Reveal Neuroplastic and Oscillatory Changes Associated with Self-Compassion and Negative Emotions. *Int J Clin & Health Psychol*, 23: 100343. https://doi.org/10.1016/j.ijchp.2022.100343

5

Aerobics, Balance, Strength, and Flexibility

~

Shall we dance? Yes! Because dancing often includes being aerobic, improving balance, strengthening some muscles, and might increase flexibility.

> **Figure 5.1**
> This illustration by David Yu, MD, of brain cells dancing captures one delightful way to earn the best of exercise as well as touch and talk for the health of it.

DOI: 10.4324/9781003462354-5

As the neurons dancing in Figure 5.1 illustrate, dancing with someone adds the very positive brain, social, and lifespan health benefits from touch that are described often in this writing.

The chapter on neurobics covers the many neuroplastic purchases of Music and Dance. This chapter will expand on the types, reasons, and benefits of being aerobic, of doing activities that improve balance, the power of strengthening exercises, and of the wisdom of increasing physical flexibility. This is among the many ways we can influence the Flynn effect[1–3] of increasing human intelligence.

When I saw the poster for the Odyssey 2000 (O2K) bicycle trek, my motivation was to go around the world and be in 45 countries, but I was not sure I could bicycle 80 miles a day. Fortunately, I was in physical therapy with a gentleman who said: "Joyce! It's only training!" I was the third person who signed up for the O2K. Unfortunately, I did not train enough. That O2K bicycle trek seriously increased my neurobics. Half of that year we stayed in our tents. That O2K bicycle trek was more strenuous than any of the hard physical labor of the farm of my youth so it also seriously increased my aerobics, balance, strength, and flexibility."

It also gave me special appreciation for how centenarians, supercentenarians, and celebrating senior successes got a serious boost with the performance of Robert Marchand!! In February 2012, this gentleman set a world record for individuals over the age of 100 by track cycling 24.250 km in one hour. Fortunately, he also cooperated with researchers to complete a two-year training program.[4] *Then, he broke his own record by being 11% faster.* In January 2014, at the age of 103, he set another world record with his indoor track cycling speed of 26.927 km in one hour. *His maximal oxygen consumption (VO2max) increased 13%* to the same range that would classify men aged 42 to 61 as physically fit.

It is *highly* significant that this is the *first time ever* that research has shown that physical performance and maximal oxygen consumption can still be increased *in a centenarian* who will do the training. Along with other "Olympic" centenarians, the message is clear that it is never too late for beginning and benefiting from exercise.[5,6]

That is huge. Let it empower and motivate ALL of us to do aerobic exercises, strengthening exercises, activities to improve balance, and increase our flexibility. In Washington state USA, factors associated with living to 100 included living in a neighborhood that was walkable.[7]

Despite all the science behind the many reasons for regular exercise, some of us fail to do that. My struggle was getting to an aerobic pace as often as research indicated would enhance my neuroplasticity. I thought putting an exercise machine in my bedroom would help but noticed I didn't get on it. So, the change needed was to roll over, get out of bed in the morning, go to the bathroom, get on the machine, and wake up later. That worked!! While being aerobic, I multitask with reading, meditating, or planning for what is ahead. Being aerobic first thing in the morning increases my energy, clarity of thought, and mood so much that you do *not* what to come between me and being aerobic on a regular basis four or more days each week.

There are some sobering findings related to sedentary behaviors. For example, questionnaire data was collected when 501,376 dementia-free individuals were recruited into the UK Databank.[8] The data included "physical (i.e., physical activity at leisure time, housework-related activity, and transportation) and mental (i.e., intelligence, social contact, and use of electronic device) activity." With an average follow up of 10.66 years, there were 5,185 cases of dementia identified. Lower risk of dementia was related to "a higher level of adherence to activity patterns related to frequent vigorous and other exercises (hazard ratio 0.65, 95% CI 0.59–0.71), housework-related activity (0.79, 0.72–0.85), and friend/family visit (0.85, 0.75–0.96)." This highlights the potential of mental and physical activities to reduce the risk of dementia. Other mental activities are discussed in the chapter on neurobics. Although the article by Zhu[8] and colleagues covers an array of physical activities, there are several others covered herein.

Since physical exercise signals the endocrine system, the term "exerkines" has been proposed for the many factors all organs secrete in response.[9] Mitochondrial function in the brain is influenced by exerkines that adipose tissues, liver, and skeletal muscles secrete. This is among the invaluable benefits of exercising because mitochondria actively regulate metabolism of energy in the neurons and other body cells, repair of brain cells, neurotransmission, and brain health maintenance. Thus, "exerkines may act via impacting brain mitochondria to improve brain function and disease resistance."

Exerkines activate synaptic pathways and might strengthen memory.[10] Brain and blood levels of exerkines change with aerobic versus resistance exercise and whether the exercise session is chronic or acute.

Irisin is a hormone that is increased by aerobic exercise. It has a crucial role in regulating the cognitive benefits of exercise and may be a "potential therapeutic for treating cognitive disorders including" Alzheimer's disease.[10] In a three-dimensional cell cultural model of Alzheimer's disease, exercise-induced irisin led to clearance of amyloid-beta, attenuated amyloid-beta pathology, "suggesting a new target pathway for therapies aimed at the prevention and treatment of" Alzheimer's disease.[11]

Physical activity was assessed by self-report in 12,201 men aged 65–68 over a three-year period.[12] Men reporting "vigorous physical activity" of 150 minutes or more each week were deemed physically active. Healthy aging included good cognition and "no major difficulty in any instrumental or basic activity of daily living." The 16.9% who were physically active had a lower death rate and higher probability of achieving healthy aging. These researchers opined that vigorous physical activity "should be encouraged when safe and feasible."

Children who were included in physical activity after school were evaluated for their "brain network modularity."[13] Modularity describes how networks functioning in your brain become divided into subnetworks with higher modularity including more connections with nodes and less connections between these modules resulting in improvements in neurocognitive efficiency. The Fitness Improves Thinking in Kids (FITKids2) program involved kids aged 7–9 years old in a physical activity for two hours after school each day for 150 days. Their multimodal intervention of aerobic and resistance exercise programs gradually increased in intensity and *exceeded* "the national physical activity guideline of 60+ minutes of moderate to vigorous physical activity per day." They found that children with the higher brain modularity at baseline showed benefits from the physical activity program of FITKids2 in improved executive functions, efficiency of cognition, and mathematical skills. The children with higher brain modularity who were not included in the physical interventions after school did not have these same cognitive and academic gains. Thus, the building of structural and functional brain benefits across activities could facilitate increasing the benefits of subsequent multimodal interventions that include cognitive stimulation.

Single acute exercise sessions and chronic exercise stimulate production of exerkines, proteins that influence cellular and molecular activities in the brain.[10] Both aerobic and resistance training stimulate this production. Personalized exercise programs can consider these factors. Improvements beginning in childhood can increase cognitive benefits in midlife.[14] This is especially important because frailty increases the risk of dementia.[15] Physical exercise influences duration and quality of sleep, cognitive function, and clearance of Alzheimer's disease biomarkers as measured by positron emission tomography.[16]

With artificial intelligence capacity to increase the depth and breadth of scientific discoveries, learn from human creative problem solving, and integrate evolving treatments, the benefits could include having more time for human influence on progress.[17] This will necessarily include psychological science applications in all categories of lifestyle choices.

Aerobic Exercise

Aerobic exercise has been associated with increased neurogenesis in humans.[18] It also influences the survival and maturation of adult born neurons.[19] In elders without known cognitive impairment exercise can improve cognitive performance.[20] Being aerobic for up to 60 minutes can improve information processing. A high level of physical fitness can decrease the so-called "normal age-related" atrophy of the hippocampus and increase volume of the hippocampus.[21] With elders whose average age was 83, it predicted greater integrity of microstructures in brain networks related to memory.[22] It can result in greater health of brain white matter in people aged 60–78.[23] Exercise energizes motor responses to improve the speed of reaction.[24] In community dwellers between the ages of 55 and 80, physical exercise was a powerful method to increase gray matter volume in the hippocampus and prefrontal cortex, "effectively reversing age-related loss in volume by 1 to 2 years" with related improvements in memory performance.[25]

An aerobic pace was frequent on the O2K bicycle trek. The goal of 80 miles a day was strenuous enough. However, the elevation gain of the many mountain passes was my biggest hurdle as I have never been good at going uphill. At first, the group I rode with would get to the top of the mountain pass and wait for me; then I would get to the base on the other side and wait for them. I suggested that we were maximizing each other's weaknesses, and it might be better if we meet in camp at end of day. First ones in could set up tents. That enabled each of us to get our aerobics, balance, strengthening, and flexibility at our best pace.

Children participated in weekly exergame sessions at "65%-HRmax" with "high cognitive challenge level (adapted to the individual performance)."[26] After session durations of 5, 10, 15, or 20 minutes, they had "an Attention Network Task." Exercise duration had a significant effect on reaction times but not on accuracy of response. After 15 minutes of exercise, their information processing speed was faster compared to after 10 minutes of the exergame. These researchers opined that "an acute 15min cognitively high-challenging bout of physical exercise enhances allocable resources, which in turn facilitate information processing, and — for more active children only — also executive processes."

Students aged 9- to 15-year-old "with better aerobic fitness can present better-processed elements and smaller omission errors."[27] There were 187 students in this study.

Being aerobic on a regular basis across the lifespan is associated with both better cognitive functioning as well as better cerebrovascular functioning in older adults.[28] At ages seven to fifteen 1,244 youth were assessed for waist-to-hip ratio and physical fitness.[14] At an average age of 44.4 years their poor "muscular endurance and power" as well as poor cardiorespiratory fitness in youth were associated with "lower midlife psychomotor attention" as well as "lower global cognition." These researchers opined that strategies in childhood to improve physical fitness and reduce obesity are important for improving midlife cognition.

After only six months of aerobic activity, women between the ages of 55 and 85 with mild cognitive losses had improved executive function.[29] Benefits of this potent nonpharmacological intervention included executive control processes of multitasking, cognitive flexibility, information processing efficiency, and selective attention. Fortunately, Laura Baker and colleagues had these women maintain the aerobic pace on cardio machines since the study took place in Seattle where the sun can seem like a foreign visitor, blind the eyes of the unaccustomed, or even send them to the malls or to the mountains.

Of course, the place is so scenic, that in rain or shine you will find a lot of folks on one of the many bicycle paths in King County … or biking the 200 miles from Seattle to Portland in one or two days, the invigorating and very rewarding STP … having completed the STP seven times, I still view it as one of my favorite rides. In addition to scenery, the other riders form a supportive community. For example, on one occasion when I was preparing to stop by the side of the road, a fellow passing asked if I was OK; as soon as he heard the issue I wanted to address, he was off his bike, fixed the problem in mere moments, and pedaled off vigorously leaving me in a pool of gratitude that I could resume the ride. Wow!!

In her research, Baker and colleagues also found that moderate to high-intensity aerobic exercise was associated with improved brain function as well as lower levels of amyloid protein in blood plasma. The amyloid protein is the major component of the plaques that are a hallmark lesion of Alzheimer's disease. At the Alzheimer's Association International Conference in 2015 Baker reported on her recent study which measured a group that stayed below 35% of their maximum heart rate during a routine of stretching.[30] They compared that group to an aerobic exercise group that stayed at 70–80% of their maximum heart rate for 45 minutes a day, four days a week for 6 months at local YMCAs. The aerobic individuals significantly reduced their cerebrospinal fluid levels of tau; they significantly increased blood flow in parts of their brain that handle memory and information processing; and these changes were related to better executive functions of attention, planning, and organizing. That is potentially huge!! That suggests that the lifestyle choice of aerobic exercise in the 70 to 80% maximum heart range could offer greater protection from Alzheimer's related brain pathology than any currently available prescription, according to Baker.

Please note also, the high probability that the only side effects from being aerobic could be improved physical, mental, brain, and emotional health. Also, appreciate that artificial intelligence can provide superior data on time, duration, intensity, cardiovascular variables, neurocognitive changes, etc., to fine tune your aerobic exercise routines across a vigorous longevity.

Note that 45 minutes 4 days a week at 70–80% maximum heart range was the aerobic exercise intensity, duration, and frequency in Baker's study and remember to confer with your healthcare provider on what is best for you. Among the reasons this is imperative is that being an endurance athlete is a risk factor for atrial fibrillation (AFib). There are studies showing reversal of AFib in patients with nonpermanent AFib.[31] However, safe is better than sorry. So, keep your healthcare provider informed about the parameters of your fitness efforts: type, intensity, duration, frequency, changes in heart rate in the aerobic pace, changes in resting heart rate, etc., so you earn the best health for your efforts.

Physical exercise has been shown to be a low cost, portable, fun, low tech way to enhance and maintain Brain Power *AT ANY AGE*®. Artificial intelligence can refine this database. How valuable could this dual purchase of Brain Health and general health be?

A review of random controlled trials suggested that physical exercise could be a powerful way to increase gray brain matter in elders such that cognitive losses and behavioral problems associated with brain atrophy can be prevented.[32] Women over the age of 60 participating in virtual scenario exergaming also showed increased brain-derived neurotrophic factor (BDNF) after a single session and improved inflammatory markers.[33] Their cognitive performance measured by the Montreal Cognitive Assessment (MoCA) found significant improvement. They opined that "exergaming might act with therapeutic potential capable of inducing neuroplasticity through BDNF modulation in this population."

Aerobic exercise might prevent MCI and be neuroprotective by stimulating nitric oxide and activating cells to reduce inflammation and enhance synaptic plasticity.[34] It might enhance functioning in individuals with MCI.

A review of 29 random control trials with 2,458 cognitively impaired adults found that high-intensity, short-duration, multicomponent, and high frequency exercise routines might "be the most effective type of exercise" for improving global cognitive functioning in cognitively impaired adults.[35] Adults in these studies were over 50 years old.

In their descriptive overview of studies researchers "highlight the neurobiological effects of physical exercise, which is able to promote neuroplasticity and neuroprotection by acting at the cytokine and hormonal level, and the consequent positive clinical effects on patients suffering from cognitive impairment."[36] Exercise has increasing evidence "as a primary prevention intervention, a nonpharmacological therapy and a rehabilitation tool for improving cognitive functions in neurodegenerative diseases."

A review of studies which used different interventions and a variety of measurements found that resistance training and aerobic exercise in older adults can significantly improve cognitive functioning as measured by the

> Mini-Mental State Examination (MD 2.76; 95% CI 2.52 to 3.00), the Montreal Cognitive Assessment (MD 2.64; 95% CI 2.33 to 2.94), the Wechsler Adult Intelligence Scale (MD 2.86; 95% CI 2.25 to 3.47), the Wechsler Memory Scale (MD 9.33; 95% CI 7.12 to 11.54), the Wisconsin Card Sorting Test (MD 5.31; 95% CI 1.20 to 9.43), the Trail Making Tests (MD −8.94; 95% CI −9.81 to −8.07), and the Stroop Color and Word Test (MD −5.20; 95% CI −7.89 to −2.51).

Since cognition improved in all elders after physical exercise, their "meta-analysis recommended that patients perform at least moderate-intensity aerobic exercise and resistance exercise on as many days as possible in the week."[37]

A single episode of aerobic exercise in healthy adults improved their learning of a motor sequence and enhanced motor cortex excitability.[38] This was a small study with 20 healthy adults with an average age of 26.4 years. They did 5 minutes of warm up exercise prior to and 5 minutes of cool down after 20 minutes exercising at target heart range. Enhanced brain connectivity after a 12-week walking exercise was associated with improved memory in people with MCI and an average age of 78 years.[39]

Reviewers of 62 reports on research studies opined that "regular exercise appears to be the most effective nonpharmacological prophylactic/therapeutic strategy" for healthy aging of vascular health.[40] Participants in these studies had an average age of 45. Combining aerobic, stretching, and resistance training improved pulse wave velocity, the "gold-standard method for arterial stiffness evaluation," and flow-mediated dilation. With older adults, stretching improved vascular function. A review of 38 studies with 2,154,818 participants found that physical exercise reduced risk of all-cause dementia including vascular dementia.[41]

MRI measured brain volumes and diffusion tensor imaging quantified white brain matter microstructure in 4,365 individuals from the Rotterdam Study.[42] Their average age was 64 years. More involvement in sports at follow up was associated with greater volume of white brain matter, of gray brain matter, and of the total brain. Since more intact white brain matter microstructure and greater brain volumes were associated with people still being more physically active when seen for follow up, people with advanced brain aging are a potential target population for novel evidence-based prevention and intervention strategies.

Exercise also helps protect the myelin sheath from degeneration.[43] The myelin sheath, primarily formed of lipids, provides metabolic support to the axon of nerve cells and increases efficiency of data transmission. Despite the need for more research to clarify type, timing, intensity, and duration of physical exercise that could maximize myelin sheath health across the lifespan, current evidence suggests that physical exercise does help maintain myelin health.

Global cognition and executive functioning were improved by physical exercise in 646 frail and impaired adults.[44] Twenty-four weeks of exercise was also associated with increased blood flow to the brain.

Exercise increases BDNF which regulates regeneration of muscle and strengthens transmission of data at the synapses.[45] By supporting this muscle-brain crosstalk, BDNF and other myokines (proteins produced when muscles contract) enhance global cognition and more complex cognitive functions such as memory, learning, and motor coordination.

A review of 100 studies over 20 years of research literature noted that both aerobic exercises and resistance training are effective in reducing depression.[46] BDNF levels are elevated in clinical populations and healthy humans with "acute exercise, particularly high-intensity exercise;" they note that acute exercise forms increased BDNF faster than antidepressants in people experiencing depression. They opine that the

> message to take home is that clinicians should consider encouraging exercise in all types of depression, following a clinical assessment of the patient's ability and willingness to exercise. They should work closely with exercise scientists to overcome the challenge of optimum exercise mode, intensity, and duration to be prescribed.

They've issued "a wake-up call" to clinicians, academicians, governing bodies, and policy makers to consider how replacing medication with exercise as therapy could significantly increase health outcomes while it significantly reduces healthcare costs. They advocate for the creation of specific courses in the health sciences curriculum. They recommend medical research trials to assess differences between antidepressants versus evidence-based exercise intervention impact on neurocognition.

Survey data collected in 2011–2013 on 13,773 community-dwelling people 40–74 years old found that both non-leisure and leisure-time physical activity were very effective in reducing risk of dementia.[47] Non-leisure activities included housework, occupational work, and commuting.

Seventy-six adults with an average age of 72.9 were in a 26-week trial of supervised aerobic exercise.[48] The active control group did non-aerobic exercises of stretching and toning. The aerobic group had increased cardiorespiratory fitness which correlated with positive changes in bilateral volumes of their hippocampi and with improvements in their memory. The aerobic group also had increased functional ability compared to the stretching and toning control group. The cardiorespiratory gains of the aerobic group were not significantly greater than those found in the stretching and toning group.

At age 46, 4,481 individuals consented to wearing an accelerometer and be tested for executive functioning and memory.[49] Compared to sedentary behavior, vigorous and moderate physical activity was associated with the best midlife cognition.

The COSMIC Collaboration looked at gait speeds of adults older than 65 on three continents in six countries to determine what their gait speeds correlated with.[50] Faster gait speeds in all cohorts in Africa, Asia, and Australia was correlated with better cognitive functioning and lower body mass index.

A literature review explored the therapeutic potential of exercise for depression.[51] It found studies showing physical exercise influencing change to a more positive mood status with structural and functional changes in the brain; increased blood flow to the brain; reduced ACTH and cortisol; increases in beneficial gut microbiota; and increased myokines and fitness levels. Thus, the studies found anxiolytic and antidepressant effects by "counteracting the complex and multifaceted pathophysiology of depressive disorders." More studies with human subjects are needed to clarify type, dose, intensity, duration, and frequency of exercise (Figure 5.2).

A lot of research has found improved brain performance associated with increased brain volume after vigorous exercise. That is particularly significant considering the so-called "normal" shrinkage that is observed in the human brain with aging. This important study by Erickson[25] and colleagues found

Figure 5.2
Join me in gratitude to Arthur F. Kramer for permission to share this graphic (from Erickson). Ascending lines show 2% increase in brain volume with aerobic exercise which includes improved memory! Descending lines of 2% in one year included decreased memory.

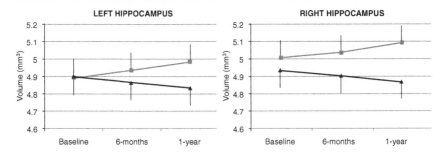

Source: From Erickson[25] (used with permission)

Note: Ascending lines show 2% increase brain volume with aerobic exercise; Descending Lines represent the so-called "normal age-related loss" of brain volume.

2% *increase* (ascending lines) in volume of the hippocampus (the part of your brain involved in memory) which was associated with aerobic exercise; compared to the 2% decrease (red lines) in brain volume that is the so-called normal brain atrophy.

Is that a significant purchase of 2% in brain volume or a significant gain of 4% considering the so-called "normal" age-related shrinkage? You decide the math.

PLEASE NOTE that this increased volume in the hippocampus in individuals who did aerobic exercise was associated with improved memory!! Is this potential for increased volume of your hippocampus, and associated improved memory, a gift you want to earn for yourself? Is this a purchase in brain health and brain power that you want to role model and to teach in academic and corporate settings knowing that it might increase health and productivity?

You do NOT want your brain volume to "normal" for your age!!

I have had a long career as an expert in medical and psychiatric matters for a variety of court systems. Expert witnesses serve the trier of fact on issues that go beyond the expertise of the court. In my role as court expert, it has often been necessary to testify about noninvasive brain studies. It has never been a source of comfort to the patient, family, friends, or to any other listener in the courtroom, to hear that the patient's brain "shows the normal atrophy of aging." Hearing about "brain atrophy" can be particularly painful to individuals with some awareness of increasing neuroscience evidence that brain shrinkage can be avoided, slowed, and reversed as studies referenced herein indicate.

Reframed in Positive Psychology:
You want your brain volume to be greater than normal for your age!!

Resistance Training

A mini review[52] found that recent research "suggested improved quality of life was associated with" resistance exercises that strengthened muscles, improved balance, and increased functional capacity and autonomy. They opined that preclinical and clinical evidence finds that resistance exercise can be part of evidence-based treatment to alleviate Alzheimer's disease symptoms.

In their systematic review of eighteen studies,[53] brain changes in response to resistance exercises were substantial "especially in the frontal lobe." These changes were associated with improved executive functioning. There was *less white brain matter atrophy* and "smaller white matter lesion volumes" with resistance exercise training. More research on this is needed to clarify dose and timing.

Participants 55 years old or older who were dwelling in the community and had been diagnosed with mild cognitive impairment (MCI) were supervised for 26 weeks in sessions two or three times a week in high-intensity resistance training that gradually increased to 80% of peak capacity.[54] This "led to better global cognitive and executive functioning at 18 months" which was associated with *increased volume in specific brain locations* as well as strengthened connections between various brain locations. "Six months of high-intensity resistance exercise is capable of not only promoting better cognition in those with MCI, but also protecting AD-vulnerable hippocampal subfields from degeneration for at least 12 months post-intervention."

Resistance exercises increase neurogenesis. Since these neuroprotective effects are "direct, delayed, and protracted" independent of physical activity after formal training ended, this research captures some of the *huge* benefits of doing strengthening exercises on pneumatic resistance machines with gradual increasing resistance to 80% peak capacity two or three times a week. Additional benefits which have been associated with resistance exercises include, but might not be limited to, effectively counteracting insulin resistance, reducing sarcopenia (weak muscles), and stimulating osteogenic mechanisms which strengthen your bones. These could reduce your risk of falling as well as reducing the risk for injury in the event of a fall.

The "direct, delayed, and protracted" benefits of resistance exercise motivate me to keep them in my morning exercise routine. However, since reading and writing neuroplasticity research is a joy while being aerobic on a machine, that is my preference. I will be grateful after a device can monitor my resistance exercises so I can at least read while doing those. Meanwhile, since both aerobic and resistance exercises increase neurogenesis, this gift I give myself and I invite you to do same for yourself.

Another benefit to resistance training is that muscle tissue stores the glycogen we might need for any race across the savannah. Thus, muscle building might be compared to the remodeling you do when you add shelves in the closet where you store nutrients to hold more canned goods.

Neuroprotective benefits of resistance exercises include the anti-inflammatory effects, reducing the deposits of beta-amyloid plaques, and increasing neurotrophic factors.[52] Better memory, improved neurocognitive functions, better balance, increased autonomy and capacity to function combine to improve quality of life. Based on their mini review of studies, these researchers opined that resistance exercises "can be proposed for patients with AD, as an alternative and adjuvant therapy, as a possible therapeutic strategy, not only to improve symptoms, but also to prevent or control the progression of neurodegeneration in" Alzheimer's disease and "even prevent AD and other types of dementia" since resistance exercise "seems to have preventive potential, alone or in combination with other types" of exercise.

A small study with 38 individuals who had Alzheimer's disease and cognitive deficits measured their handgrip in nondominant and dominant hands.[55] MRIs were done on these adults. Greater strength in their nondominant handgrip was associated with higher brain volume in their hippocampus. Greater dominant handgrip was associated with greater volume in their frontal lobes. Higher scores on the 2 Minute Walk Test were associated with larger hippocampal, frontal, temporal, parietal and occipital lobe volumes.

Again, in Positive Psychology:
You want your brain volume to be greater than normal for your age!!

Data from the US National Health and Nutrition Examination Survey (NHANES) on 777 people indicated better verbal learning and verbal fluency in nondemented US men over age 60 who had maximal handgrip strength.[56] There was a prominent dose response "with no indication of a plateau."

Tai Chi

A small study with 36 healthy adults obtained fMRI and measurements of behaviors before and after Tai Chi training.[57] Compared to general aerobic exercise,

Tai Chi Chuan significantly enhanced the nodal clustering coefficient of the bilateral olfactory cortex and left thalamus, significantly reduced the nodal clustering coefficient of the left inferior temporal gyrus, significantly improved the nodal efficiency of the right precuneus and bilateral posterior cingulate gyrus, and significantly improved the nodal local efficiency of the left thalamus and right olfactory cortex. They also became more cognitively flexible after the eight weeks of three sessions weekly.

Six individuals 55 or older enrolled in a Tai Chi program of twelve weeks.[58] Both brain and muscle magnetic resonance spectroscopy (MRS) were obtained before and after training. They opined that their "results suggest that Tai Chi, as a mind-body exercise, may effectively promote neuroplasticity and increase lower extremity muscle oxidative capacity in older adults." This supports the psychological and physical benefits of Tai Chi.

MRIs and cognitive testing were done with 42 healthy women with an average age of 62.9 who either did Tai Chi or a brisk walking exercise.[59] All women had a Montreal Cognitive Assessment (MOCA) of 26 or above indicating good intelligence. Those in the Tai Chi group had previously had six or more years of "Yang-style Tai Chi" training with five 90-minute sessions weekly at minimum and a "Yang-style Instruction Certificate." More than six years of at least five weekly sessions of 90 minutes walking exercise qualified women for the walking group. The women who did Tai Chi had episodic memory that was superior to that in women who walked briskly. Women in the Tai Chi group also had *Higher gray matter density* in the inferior and medial temporal regions" *of their brain including the hippocampus.* These researchers opined "that long-term Tai Chi practice may improve memory performance via remodeling the structure and function of the hippocampus." Long-term practice of Tai Chi is associated with faster reaction times and greater accuracy of response.[60]

With a small sample of healthy adults aged 50–60 functional near-infrared spectroscopy (fNIRS) assessments were used to test "oxyhemoglobin changes in the prefrontal cortex (PFC), motor cortex (MC), and occipital cortex (OC)."[61] MOCA measured global cognition. One group had at least three years of Tai Chi experience. The other group had no history of yoga or regular activity. MOCA scores were significantly higher in the adults with Tai Chi experience compared to adults without Tai Chi experience. Their visual spatial and memory skills were significantly better. Their higher global cognitive abilities compared to the group not using Tai Chi "may be associated with higher brain functional connectivity of the" prefrontal cortex, motor cortex, and occipital cortex.

Tai Chi lessons of one hour twice a week for 16 weeks were given to 160 community-dwelling adults.[62,63] Results included cognitive benefits and improved independent activities of daily living. To improve cognition Tai Ji Quan training delivered remotely via videoconferencing as a training intervention for elders with MCI living in the community is feasible, acceptable, and safe.[64]

Neural function components of working memory were measured using ERP-fNIRS in healthy individuals aged 65 or older.[65] The Tai Chi intervention was three days a week for twelve weeks with five-minute warm up and cool down and 50 minutes of "Eight Forms Five Steps" with moderate exercise intensity. The control group maintained their normal lifestyle. The Tai Chi group had significantly faster reaction times than the control group. They also demonstrated more neural activity. Individuals using Tai Chi had more memory improvements than the fast-walking group. Since Tai Chi is a mind-body intervention which integrates aerobic exercise and cognitive training, this could explain the better memory functioning in adult Tai Chi practitioners. They opine that "Tai Chi, a simple and low-risk mind-body exercise, can effectively alleviate cognitive decline in the elderly."

Another small study[66] found significantly greater connectivity increases with Tai Chi compared to health and wellness education. Connectivity increases in both groups were mainly in the default mode network.

Taekwondo

Muscle and joint movements in the whole body are stimulated by many and different taekwondo movements.[67] This influences growth factors release, muscle strength, and cardiorespiratory fitness. In children physical exercise promotes growth with the increase in growth hormone and increased insulin-like growth factor-1. In their review of twelve random controlled trials with 260 children and adolescents aged between 8 and 16 years old, taekwondo practice increased their secretion of growth hormones compared to students not doing taekwondo. Taekwondo has also been associated with increasing neurotransmitters.

In a small study with fifteen 13-year-olds who had never used taekwondo these adolescents were assessed by resting MRI, self-report, and Body Intelligence Scale.[68] Compared to the control group, after taekwondo training, these children "showed increased functional connectivity from cerebellum seed to the right inferior frontal gyrus." These researchers concluded that "Taekwondo training improved body intelligence and brain connectivity from the cerebellum to the parietal and frontal cortex."

Diffusion tensor imaging with 30 taekwondo athletes found *"increase in total brain volume, gray matter, frontal lobe and precentral gyrus volume"* associated with the taekwondo training in these athletes compared to the control group.[69] These researchers concluded that the "increase in the volume of gray matter, frontal lobe, postcentral gyrus and corticospinal tract together with the brain volume shows that taekwondo exercise contributes to physical, spiritual and mental development."

In a small study with children given taekwondo training five times a week for 16 weeks, their self-confidence and cognitive tests improved significantly leading researchers to suggest that "regular taekwondo training may be effective for enhancing cognitive function and academic self-efficacy in growing children."[70] They opined that their improved cognitive function was related to higher levels of growth factors related to enhancing neuroplasticity.[71] In a similar study protocol with women 65 years old and older they again found improved cognition with taekwondo training which was thought to be related to "increased neurotrophic growth factor levels."[72]

Adolescents with attention deficit hyperactivity disorder had better selective attention after a year and a half of taekwondo training.[73] Since one year of taekwondo that was age-adapted for people over 40 years resulted in come cognitive gains, researchers opined that taekwondo "may offer a cheap, safe, and enjoyable way to mitigate age-related cognitive decline."[74]

HIIT

Youth with an average age of 12.83 did five-minute warm up exercises prior to a supervised session of 16 minutes of HIIT.[75] Each of the four sets of 1-minute HIIT exercise included 30 seconds of work and 30 seconds of rest. The matched control group watched a documentary. These HIIT sessions had a significant "positive effect on perceptual-attentional skills." With improved visuoperceptual ability and selective attention, including a "HIIT session at the beginning of the school day" could enhance academic performance.

Twenty-five male college students with an average age of 21.7 years did HIIT of ten 1-minute repetitions of cycling at VO_{2peak} with 1 minute of passive rest between these.[76] Then they sat in isolation with no interference for 30 minutes. They had no mental disorders, were not multilingual, not musicians, and not in treatment for any central nervous system issues. Their mood and Digit Span Test were assessed pre-, post-, and 30 minutes after intervention. Although their mood was significantly decreased after exercise and they were initially easily confused, their mood significantly increased 30 minutes after HIIT. Digit Span scores were significantly higher immediately after HIIT and 30 minutes after HIIT suggesting that even brief HIIT can be a strong and healthy intervention to improve mood and enhance cognition.

Healthy female university students aged 18–25 years participated in supervised HIIT at "85% of maximal heart rate" three days a week for four weeks.[77] Four aerobic repetitions were preceded and followed by five minutes of warm up and cool down; these repetitions were separated by resting ten seconds. The HIIT participants had a significant decrease in body fat percentage and waist circumference from the second week through the end of four weeks, whereas these changes were not noted in the matched control group which maintained usual physical activities with no training. The HIIT group had significant improvements in cardiovascular function, psychological well-being, and executive functions which were not noted in the matched control group.

A small study looked at the immediate effects of high-intensity functional training.[78] In these nineteen mature males who had practiced HIFT for at least four times a week and at least three years, evaluations included lactate level, repeated jumps, balance and cognitive tasks. Training included whole body exercises for 30 seconds with 30 seconds rest intervals and encouragement to use maximum effort. Cognitive tasks were evaluated for how many correct answers they could give in 30 seconds, their average response time, and the number of errors. Lactate levels increased from pretest to the third posttest with a slight decrease from the third to the sixth posttest. Significant balance improvement was only seen between the pre and first posttest. There was no significant change in the number of jumps. Correct answers significantly increased from pretest to posttests 3, 4, 5 and 6 without differences between the posttests. Significant cognitive improvements were associated with this trend for improving lactate values and "lactate derived from skeletal muscles is metabolized by neurons, with the direct consequence of enhancing executive functioning." Improved blood lactate levels with exercise enhances cognitive functioning in both accuracy and speed of task performance.

After eight weeks of HIIT intervention with older sedentary females, MRI scans, fitness as measured by VO_{2max}, and cognitive assessments were obtained.[79] The cognitive improvements that followed were not explained by brain structural changes.

A very small study with only 12 healthy young males suggested that duration of HIIT might be a factor.[64] Increasing BDNF, VEGF-A, and faster response times were more effective after 20 minutes of HIIT than after 30 minutes.

Another small study included 32 recreational tennis players with an average age of 21.4.[80] They were assessed at baseline and after HIIT with and without cognitive load via incongruous data. Their reported perceived exertion significantly increased in both conditions but did not differ. Cognitive load significantly decreased tennis serve accuracy after HIIT.

Semantic fluency and executive functioning significantly increased after HIIT compared to moderate-intensity continuous training[81] in healthy adults with an average age of 24. These cognitive improvements were associated with increased lactate levels.

Exercise intensity influences lactate production and may help alleviate some chronic diseases.[82] "Lactate is related to vascular endothelial growth factor (VEGF) and BDNF expression and can improve brain functions, such as angiogenesis and neuroplasticity and stress-related symptoms such as depression, by regulating the function of hippocampal mitochondria." Lactate can cross the blood brain barrier to increase BDNF in the hippocampus. Adult neurogenesis persists in the hippocampus.[83] This could play an important role in managing the healthcare of individuals with Alzheimer's disease.[84]

Dance

It has been a privilege to meet with the Music Mends Minds® group[85] and bring them an extensive overview of research showing the value of their use of music as medicine.[86] For World Brain Day 22 July 2023, we had activities on *four continents plus virtually.*

Cognitive processing of complex information; integrating information from sight, sound, balance, and emotion; planning and doing motor activities; adjusting your pace if seeking the benefits of being aerobic; and coordinating with other people if not dancing alone … all these factors contribute to the value of dance being among the activities of the Music Mends Minds® organization.

This organization was launched by Carol Rosenstein after her husband was diagnosed with Parkinson's; despite his increasing dementia, he had such a profoundly positive response to music, becoming more engaged and happier, that this organization has increasing numbers of 5th Dementia Band events around the world and via the internet. MusicMendsMinds.org[85] is a rich source of information on the multiple ways that music is medicine; this includes podcasts on the website.

PlayingForChange.org[87] is another rich source of information. It includes videos of an international collection of individuals creating music videos, videos of youth dancing, and discussing how these contribute to their efforts to promote peace.

Individuals with Parkinson's disease (PD) or multiple sclerosis[88] benefit by "escaping, expanding, and embracing." This helped them more readily experience their "illness as a journey" through which they could learn.

A mini review[89] noted that: "non-pharmacological interventions such as dance therapy are becoming increasingly popular as complementary therapies for PD, in addition to pharmacological treatments that are currently widely available." Their review of research "evidence suggests that dance interventions can induce neuroplastic changes in healthy older participants, leading to improvements in both motor and cognitive functions." Additional research is needed to increase our understanding of neuroplastic purchases of dance interventions across the lifespan and with the many forms of compromised brain chemistry, architecture, and performance.

A greater volume of brain white matter associated with cardiorespiratory fitness was found in overweight and obese children.[90] In the ActiveBrains project, a *greater volume of brain white matter* was related to motor fitness. Some regions of greater white matter volume were related to academic performance. Since white matter volume increases are related to cardiorespiratory fitness and academic performance, exercise interventions combining aerobic, muscular, and motor training need to be encouraged for better brain development, physical fitness, and academic success.

Dance and music could be among your best choices for enhancing neuroplasticity. They are so important to brain power that both are included in the chapter on neurobics with much more data there on hearing music and actively making music.

A review of 35 studies "showed that dance improves brain structure and function as well as physical and cognitive functions."[91] By adapting style and intensity to the physical and personal needs of the adults, dance training protective effect of neurocognitive functioning could be enhanced.

College students with expert level dance training were compared to non-dancers with equivalent education.[92] When viewing dance sequences, they activated the action observation network more than nondancers. In a balance task and video game, dancers had superior performance. Dancers had altered connectivity of the "general motor learning network" and of the action observation network. Their connectivity functional differences were related to "dance skill and balance and training-induced structural characteristics" although there was no difference in cognitive abilities and "little evidence for training-related differences in brain volume in Dancers."

After six months of supervised dance intervention three times a week in 60-minute lessons, "intervention-specific complex brain plasticity changes that were of cognitive relevance" were noted in volunteers older than 60 years who were healthy.[93] Compared to comparable individuals living life as usual, people in the dance intervention group had improved attention. They also had

> significant rs-FC increase of the default mode network (DMN) and specific internetwork pairings, including insulo-opercular and right frontoparietal/frontoparietal control networks ($p = 0.019$ and $p = 0.023$), visual and language/DMN networks ($p = 0.012$ and $p = 0.015$), and cerebellar and visual/language networks ($p = 0.015$ and $p = 0.003$).

Thus, these researchers opined that their supervised dance intervention "led to intervention-specific complex brain plasticity changes that were of cognitive relevance."

A review of fourteen random controlled trials found that dance interventions can promote mental health in healthy elders "suggesting that the multimodal enrichment tool is a potential strategy for health promotion and prevention of" Alzheimer's disease.[94] They recommend more high-quality studies of dance interventions to gather more evidence of specific functional and associated neurophysiological correlates.

A review of eight random controlled trials found that dance improved visuospatial function, memory, global cognition, and language.[95] The positive effect on quality of life was highly significant. These reviewers opined that dance is an inexpensive nonpharmacological intervention that can slow cognitive decline in individuals with mild cognitive decline. They caution that more "design studies with long-term follow-ups, neuroimaging, biological markers, and comprehensive neuropsychological assessment are required to understand the mechanism of dance interventions and demonstrate its efficacy for older adults with mild cognitive impairment."

Thirty-five individuals with an average age of 71.5 years who had amnestic MCI participated in group aerobic dancing three times each week for three months.[96] Before and after the intervention they had cognitive assessments and an MRI. *Greater right and total hippocampal volumes* were evident after the intervention as compared to the control group. The intervention and episodic memory were correlated "as the intervention group showed a higher Wechsler Memory Scale-Revised Logical Memory" score. They concluded that *"aerobic dance intervention has great potential for enhancing cognitive function by increasing hippocampus volume."*

A review of 24 studies with healthy elders and five studies of individuals with Parkinson's disease looked at dancing, treadmill training and Tai Chi to compare how these as long-term exercises regulate brain activity.[97] All studies found significant improvements after the types of exercises chosen. "An inverse change trend on the functional connectivity in people with PD was observed after treadmill training, whereas increased brain activity, cognitive function, memory, and emotion were noticed in healthy older adults."

Structural and resting-state MRI data was collected on 43 ballroom dancers who were at expert level. This same data was collected on a matched control group of nondancers.[98] Researchers "measured their empathic ability using a self-report trait empathy scale." Compared to the nondancers ballroom dancers reported greater empathic concern. The ballroom dancers "gray matter volumes in the subgenual anterior cingulate cortex" correlated with their increased empathic concern. The dancers brain connectivity between the occipital gyrus and anterior cingulate cortex was positively correlated with both the number of years with dance partners and the empathic concern scores. They opined that their "findings provided solid evidence for the close link between long-term ballroom dance training and empathy, which deepens our understanding of the neural mechanisms underlying this phenomenon."

A meta-analysis of nineteen studies indicated that "dancing can improve mobility and endurance compared to no intervention and afforded equivalent outcomes compared to other exercise programs."[99] They opined that dance interventions are safe and effective for elders in the community.

More prefrontal cortex activity and significantly improved cognitive function, measured by the Japanese version of the MOCA, after video game dance training was evident with a group of adults who had MCI.[100] Also, a "tendency toward improvement was observed in the trail making test in the mild cognitive impairment group."

A review of research on individuals over 55 years old who participated in supervised group dance interventions considered 50 published articles.[101] They noted that "most interventions were 12 wk long, 3 ×/wk, for 60 min each session. The dance styles most used were ballroom and cultural dances." The most frequently assessed functional outcome was balance. They concluded that functional gains, especially balance, can be induced in elders with any dance style.

A recent integrative review of studies[102] with healthy elders aged 59 or more found that "dance practice was associated with an improvement in functional connectivity, cognitive performance, and increased brain volumes." With regular dancing of sufficient intensity, several regions of the brain benefited from the increased use of visual, auditory, motor, and executive functions required to move in space to rhythm and to others in the environment. Improvements were noted in "cognition, attention, executive functions, and memory." Neuroplastic changes can "significantly increase and also preserve the performance of the functional capacity (postural control, walking speed) of older adults, contributing to their autonomy and quality of life."

A weekly event on the farm of my youth was going to a local club where square dancing was the major focus. There would also be a few songs played for other forms of dancing. Despite the daily dose of strenuous physical work required to hoist those hay bails and bags of grain, square dancing was a treat on a Saturday night. Aerobics, balance, and strengthening for the health of it, neurobics and FUN.

Use them and they will grow!! Fun!!
You want your brain volume to be greater than normal for your age!!

Another reason to use music and dance to enhance neuroplasticity is that they are portable, culturally adaptable, easy, FUN, and can be at little or no cost. You can add music and dance to your routine. Community musical activities with dance can be a significant way to enjoy and share these with the potential of increasing brain power for ALL. Simply hearing music people tend to move, some dance. If the dancing is aerobic, the benefits increase.

Healthy people over 60 years old were given an hour, three times a week, of a dance intervention and compared to age peers living "life as usual."[93] Irish Country, African, Greek, and other types of dances were included with medium intensity. With the dance intervention, enhanced brain activity on resting-state fMRI was associated with improved attention and executive functions. Depression in people with MCI and dementia was reduced with dancing.

Traditional dance of India improved balance and motor skills more than neuromuscular training in "children with Down syndrome."[103] Benefits of dance can be facilitated remotely.[104]

A meta-analysis of five studies found that after one to three weeks of Latin ballroom and aerobic dances with light to moderate intensity, adults with MCI had better brain function.[105] Participants had better immediate and delayed memory, attention, visuospatial ability, and global cognition. Researchers believed this might be related to changes in brain chemistry including increased BDNF and growth factors. Their dances were culturally adapted and with a partner or in groups, adding social interaction and socio-emotional factors. Limitations of these findings include the small number of studies reviewed.

A review of five random controlled trials found that dance-based interventions significantly decreased depression although follow-up data indicated diminishing effects.[106] Dance more effectively reduced depression in individuals with dementia than it did for people with MCI. The effect was better with dance sessions twice a week for an hour than it was for two to three times per week for 35 minutes.

Improved cognitive functioning and memory could be related to improved connectivity between and within brain networks.[107] Individuals with an average age of 78 who participated in a walking exercise had a resting-state fMRI and cognitive testing. Significant cardiorespiratory improvements correlated with better memory in healthy as well as in individuals with cognitive impairments may have been related to "increased within- and between-network connectivity" after the walking exercise intervention.

Aerobic exercises and resistance training influence the survival and maturation of adult born new brain cells. Both exercise types can be part of dancing and both exercises increase the amount of BDNF you produce. BDNF is essential for crucial nervous system functions including "survival, differentiation, and maturation" of your various brain cells.[108] It is neuroprotective of the brain cell that creates the BDNF; and BDNF is neighborly in sharing these benefits with your other brain cells. In a study of elders with cognitive impairment, exercise improved cognitive performance.

Research with elders whose average age was 83, found that aerobic exercise predicted greater integrity in microstructures in brain networks related to memory. Research found that it was associated with greater health of brain white matter (axons, the insulated branches off brain cell bodies that communicate to other cells) in people aged 60–78. Exercise energizes motor responses to improve the speed of reaction. MRI and BDNF serum levels were obtained on 142 adults without dementia.[109] Smaller volumes of their hippocampus and lower BDNF serum levels were associated with poorer spatial memory. They concluded that their "results identify serum BDNF as a significant factor related to hippocampal shrinkage and memory decline in late adulthood."

Thus, resistance and aerobic exercise that are part of dancing might be a key component of environmental enrichment across the lifespan for driving brain plasticity in a positive direction. In community-dwelling people between the ages of 55 and 80, physical exercise was a powerful method to *increase gray (where the brain cells are) brain matter volume in the hippocampus and prefrontal cortex "effectively reversing age-related loss of volume by 1 to 2 years" with related improvements in memory performance.*[25]

My passion for sharing new neuroscience with you includes the plan to increase our use of evidence-based interventions in hopes of making brain atrophy with aging a relic of the past. One of my objectives is to influence a revolution in human habits to increase human Brain Health and Brain Power *AT ANY AGE*® across an increasing health span for all who want vigorous longevity with maximum Brain Power. Hopefully, this can be accelerated with artificial intelligence using evidence-based interventions for increasing human intelligence in leveraging the Intelligent Health Revolution[110] in ways that can influence/hasten the Flynn effect[1-3] in *many* more countries as a way of improving global statistics on neurocognitive health and functioning. Let's work together in Hacking Neuroplasticity for the health of it!

While it is significant that researchers found increased brain volume with aerobic exercise and resistance training, it is much more noteworthy that they also report that these gains in brain volume are associated with an array of improvements in brain function and brain connectivity. This association has been found across a broad spectrum of the animal kingdom even with known compromised function prior to being aerobic.

Ongoing education such as is provided herein could increase use of this neuroscience to improve your brain health and power. However, sometimes education is not followed by action. As a psychologist and a nurse, I have assessed and provided information related to improving health parameters such as cardiovascular fitness. The usual response was something like: "Yes. I know. I gotta get on it." After providing information about the potential impact on brain health and brain power, responses were more like: "Give me the guidelines!! I'm doing this!!"

Artificial intelligence can play many beneficial roles in answering that request: "Give me the guidelines!! I'm doing this!!" It can help assess your current status to personalize intervention strategies devised during wise counsel with your healthcare provider. It can provide the unique positive reinforcement most likely to increase your compliance AND pleasure in hacking your own neuroplasticity. One reason that matters is that people are more likely to do what is FUN. In our Intelligent Neuroplasticity Revolution, we can promote much fun in aerobic, balance, strength and flexibility exercises for many people who want their brain volume to be greater than normal for their age for the overall health of it.

The Benefit List of Your Intelligent Neuroplasticity Revolution

To simplify, neuroscience research associating physical exercises to chemical, architectural, and functional brain changes for improved health suggest that you might earn these benefits:

Create many MORE new brain cells
Increase brain volume in areas involved in memory, speed, and executive functioning
Increase complexity of existing brain cells
Learn more efficiently
Have **better executive functioning**
Be more able to learn from your mistakes
Have **better memory**
Improve attention and concentration
Be faster
Improve your sleep
Use up chemistry of stress
Release frustration
Improve self-esteem
Improve well-being
Reduce depression better than prescription meds
Protect against brain damage
Save $$$$$ on medical bills
Have more energy
Possibly reduce insulin resistance

Raise **healthy** cholesterol and lower LDL
Improve your immune response
Reverse some of the effects of aging
Decrease inflammation
Increase antioxidants
Stabilize your blood pressure
Lower your resting pulse
Increase lung capacity
Avoid injury — aerobic muscle fibers resist injury better
Burn fat — aerobic muscle fibers burn fat, providing endurance and weight loss
Increase prolactin (cuddle factor and lower anxiety)
Increase Bone Mineral Density (BMD)
Maybe reduce toxins in your brain – even those associated with Alzheimer's
Aerobics, Strength, Balance AND Flexibility Exercises … essential … AT ANY AGE!!
For ALL!!

Wake Up Call (Discussion)

After reviewing 100 manuscripts that researchers had produced over 20 years, Jemni and colleagues[46] opined that their

> review could be considered a wake-up call to the policymakers to start thinking if exercise could be considered as a therapy that could potentially replace medication and save a significant health-related cost for the government. This would also imply academic institutions and the governing Ministries integrate specific courses within the medical and health science curriculum enabling future medical research.

Succinct. On point.

Let the Jemni and colleagues review, plus the tsunami of related findings (nobody can read the entire library AND include real time reports flowing in after the review), also be "a wake-up call" for all individuals seeking maximum brain, physical, social, and emotional health across their entire lifespan. Let it be "a wake-up call" for businesses that want to maximize all of these for their employees as well as appreciate how much this might increase productivity. May we also maximize the power of artificial intelligence to personalize, provide feedback on modifications based on measurements, provide positive reinforcement, prompt for scheduled exercise interventions as needed, and influence research going forward. Positive reinforcement is a key component to maximizing the benefits of exercise to every extent imaginable and devices powered by artificial intelligence can administer this personalized reinforcement of progress with great immediacy, greater accuracy, and greater effect than could ever be accomplished by any human.

All things considered it is URGENT to increase physical fitness with a variety of exercises at any age. An important point worth emphasizing is that many studies support prescribing regular and intense aerobic exercise and muscle strengthening for preventing, slowing, and potentially reversing the so-called "normal course of aging." Let us work and play and dance together in this effort to improve muscle strength; cardiovascular health; and brain chemistry, architecture, and performance. We can be role models as well as trainers to increase the number of people in our global communities that plan and change to *replace* "senior moments" with *seriously celebrating senior successes.*

Considering research data on how exercise plays a role in remyelination, researchers "prompt physicians to encourage patients to perform adapted physical activity."[111] Although it is unlikely that one treatment strategy could be sufficient, "exercise could be the key to achieving clinically meaningful functional improvement" and "moving is always better than not moving!" Artificial Intelligence could be key to achieving this more accurately and more individual-specific.

Regular physical exercise seems to have been one of the most important variables associated with successful vigorous longevity. Centenarians and supercentenarians thrived with a variety of physical activities that helped maintain their physical capacity and independence. Regular physical activity, whether in completing marathons, directing a symphony, or gardening, has been associated with greater vigorous longevity with a healthspan that could approximate the lifespan. Our centenarian and supercentenarian examples of thriving give us the clear message that it is never too late to start and to benefit from exercising!!

Aerobic exercise, dancing, muscle strengthening exercises, balance training, and stretching for flexibility are non-pharm prescriptions with so many positive side effects that I tend to cover this in as much detail as possible in every conceivable opportunity *AND* update this information as new neuroscience research results come in. There is no good reason to wait for additional studies to establish the ideal dose, timing, type of exercise, and gender differences in response. There is every reason to work with your healthcare provider to tailor the various components of your fitness routine to your own specific needs while remembering that a variety of types of physical activity and a high level of physical fitness can decrease the so-called "normal age-related cognitive decline" (ARCD) and brain atrophy (shrinkage). It could be possible, instead, to increase the volume, enhance the architecture, improve the performance, change brain chemistry, and maximize the health of your brain cells in the hippocampus (the part of the brain associated with memory and learning), in your prefrontal cortex (that plays a key role in executive functions), in connections between various parts of your elegant and complex brain, and of your Brain Power *AT ANY AGE*®. Marian Diamond[112] was prescient when she predicted that *"Enriching Heredity"* is possible across a long human lifespan.

Between their sixth and 8th week of age, when new brain cells are young and excitable, we can have the dual impact of increasing the number of new brain cells that survive AND influencing their career choice. That is another way of saying that these new cells that are young and excitable require experience such as physical activity and complex new learning to survive and to become integrated, stable, and best suited to handle the tasks that could advance your social, emotional, intellectual, career and business performance. That's among the reasons that Music and Dance might be your favorite form of exercise, especially if your dancing is aerobic and includes the social benefits of dancing with others. More on that in the chapters on neurobics and love. Increasingly, researchers are recommending exercise as a prescription for brain health for reasons referenced above, for *"Enriching Heredity,"* and to protect your brain from injuries related to falling.

Without physical exercise prescriptions, the harms that can result are similar to those of smoking and obesity. In contrast, it is unlikely that the benefits of physical activity would be overstated. You have already learned a great deal about how much aerobics can improve your heart health. It also increases blood flow to your brain and many body parts. It improves the health of your blood vessels. It can reinvigorate your immune system. Exercise can be as effective as prescribed medications for treating depression. Whether aerobic or resistance training, it is important to gradually increase the difficulty, competing against your own personal best, to improve muscle mass, muscle strength and/or cardiovascular capacity.

Most exciting to me and to audiences I have informed are the many ways exercise can improve brain health. Physical fitness beginning in childhood influences structural and functional neurodevelopment which could be highly beneficial for healthy aging across the lifespan.

All things considered it is URGENT to prescribe exercise at any age because "fitness training can also enhance cognitive vitality of older adults"[113] and because it can be considered beneficial for body and brain.[11,16,18,22,23,25,32,26,114–119]

Children in after school physical activities had improved executive functions, efficiency of cognition, and mathematical skills.[13] This could facilitate increasing benefits of subsequent multimodal interventions that include cognitive stimulation. Weekly exergame sessions facilitated information processing for children[26] with more active children also improving executive processes. Teenagers with better aerobic fitness had improved cognition.[27]

Aerobic exercise and resistance training build brain power in humans by increasing neurogenesis (birth of new brain cells) AND by being neuroprotective to preserve your brain health and cognitive functions. In the Rotterdam Study of 4365 middle-aged and older adults, MRI studies found that participants with a high level of[42] sports activity had greater volume of gray matter, white matter, and total brain volume. Neuroplasticity improved with low as well as with high-intensity exercise.[120] Benefits were also noted with higher levels of physical activity during leisure time.[47] Maintaining cardiorespiratory fitness assessed with noninvasive studies in people aged 58–81 years was emphasized in finding associations of fitness with white and grey brain matter.[121] Faster reaction time was found after high-intensity interval exercise than after moderate-intensity continuous exercise.[119] With only one session of aerobic exercise, 20 healthy people learned better and had "enhanced cortical excitability."[38] With only 20 minutes of high-intensity interval training, 12 young men had improved cognitive functioning and higher "serum levels of BDNF and VEGF-A" which are associated with brain and blood vessel health. Since elite athletes with aerobic or high-intensity interval training had better cognitive flexibility and attention than people lacking these trainings, lifespan cognitive development could benefit from sports related activities.[122]

Intensive training and practice of musicians and athletes has been associated with "use-dependent plasticity that may be subserved by long-term potentiation" processes.[118] Before and after "repetitive transcranial magnetic stimulation" excitability of their motor cortex ws evaluated and compared to a control group that was neither athletic nor musicians. Motor evoked potentials were strongly facilitated by the "plasticity-inducing protocol" in musicians and in athletes but were weaker in non-musicians and non-athletic participants. These researchers wrote that their "findings may explain one factor contributing to the higher inter-individual variability found with" motor evoked potential data. "Practice makes plasticity" and this greater plasticity capacity implies how to influence "learning paradigms, such as psychotherapy and rehabilitation" which could include "recovery from neurological/mental disorders."

Combining low-intensity resistance exercises and aerobic exercise to reduce cellular lipid peroxidation is recommended.[123] These researchers opined that with enough resistance and aerobic exercise antioxidant supplements might not be needed.

Functional near-red spectroscopy was obtained on young athletes[124] at different exercise intensities. These stages were initial, intermediate and sprinting in race-walking. In the initial stage functional connectivity was primarily between the motor cortex and the occipital cortex. The intermediate stage included connectivity with the prefrontal cortex. During the sprint all three of these brain areas were included in functional connectivity with significant changes in the prefrontal cortex and motor cortex. They opined that the more extensive activation of the brain with the greater intensity of race-walking "could facilitate the integration of neural signals such as proprioception, motor control and motor planning, which may be an important factor for athletes to maintain sustained motor coordination and activity control at high intensity."

A five-year follow up found that physical inability was a powerful predictor of prognosis for Chinese centenarians.[116] Neither cognitive impairment nor depression had the strong effect on mortality that was seen with physical inability.

Diffusion tensor imaging of healthy people with an average age of 69 and adults with amnestic MCI and an average age of 72 was done to examine white brain matter integrity.[125] There were "significant interactions between cognitive status and body mass index (BMI) on diffusivity outcome measures." Memory performance was positively associated with white matter integrity. These researchers opined "that modifiable lifestyle factors may affect white matter integrity."

There is no reason to hold back on these personalized prescriptions while evolving research refines guidelines on preferred dose, timing and method. It will also be essential to develop guidelines for all ages. A commonly used formula subtracts age form 220 and uses percentages of this number to set pulse ranges during aerobic exercise. That needs to be revisited to properly accommodate centenarians so they can enhance neuroplasticity while improving their general health.

Robert Marchand is one of our inspiring models of centenarian success.[4] This elite cyclist at age 101 set a record for centenarians by cycling 24.25 km in one hour. He then cooperated with a two-year training program. At age 103 years his speed of 26.92 km in an hour was an 11% improvement in his speed. Although his lean body mass and body weight did not change, his "maximal oxygen consumption (VO_{2Max}) increased" 13%. Please note: This is the first ever documented improved "V_{O2max} and cardiorespiratory factors" in a centenarian.

In the Cardiovascular Health Study, calorie expenditure was measured along with assessments of cognitive functioning and MRI measurements of brain volume[126] in individuals with an average age of 78.3. Their findings suggest "that simply caloric expenditure, regardless of type or duration of exercise, may alone moderate neurodegeneration and even increase GM volume in structures of the brain central to cognitive functioning."

A study of 693 adults living in the community used participants' memories of their lifestyle factors at midlife and did neuropsychological testing of current functioning.[127] Among them 64 had "super-cognition" best described in three groups. Superior immediate memory was found in about half of these people. Others with "super-cognition" had superior visuospatial abilities, language and attention. "Participants with super-cognition reported less participation in social activities and, frequently, working more than 9 hours/day and feeling stressed, at midlife. Super-cognition among the elderly is associated with having a busier, more socially isolated and stressful midlife."

Given variability of research protocols and participants, caution in interpretation of data on neurocognitive gains of physical exercise is fitting.[128] More research is needed.

Artificial intelligence influenced by researchers, businesses, academicians, and consumers can expedite standardization of research methods of a wide variety of exercise interventions (duration, intensity, mode, design), subject populations, neurocognitive factors (speed of processing and accuracy before, during, and after intervention), and other research variables of interest. Artificial intelligence can expedite human research applications with very large samples of thousands of individuals with cultural diversity. The timing is urgent to develop this collaboration between the various partners in the integration of artificial intelligence to increase the rapidly evolving neuroscience database, develop personalized applications of these evidence-based interventions, provide effective positive reinforcement to increase implementations, and protect personal identity as we respond to evolving neuroscience by modifying recommendations for the use of evidence-based interventions that could enhance neuroplasticity at all ages while simultaneously improving physical, mental, emotional, and social health across the lifespan.

What You Can Do

Add Evidence-Based Activities that Can Protect and Enhance Your Brain Power

Dance ... alone or with others ... for fitness, socializing, joy, and for neurobics!!

Be aerobic (with guidance from your healthcare provider) to:

> Increase neurogenesis.
> Improve emotional and cardiovascular health.

Do strength training three to four days a week to:

> Increase neurogenesis,
>   improve your stability, and
>   build more glycogen storage capacity.

Stretch to improve your flexibility and to increase your range of motion.

Do balance training to reduce your risk of falls to protect your brain from injury.

Establish your target heart rate and maximum heart rate. Defer to the guidance of your healthcare provider. One formula for calculating your maximum heart rate is 220 minus your age. For a 50-year-old person the maximum is 220 − 50 = 170 beats per minute (bpm). Thus, the range for a 50-year-is 170 beats per minute as their maximum and:

> 50% level is 170 × 0.50 = 85 bpm;
> 60% level is 170 × 0.60 = 102 bpm;
> 70% level is 170 × 0.70 = 119 bpm; and
> 80% level is 170 × 0.80 = 136 bpm.

Substitute your age in this example and do the math to determine your target heart rate and maximum heart rate that this formula suggests.

It is supremely obvious here that conferring with your healthcare provider is essential. For example, 220 minus the age of our oldest old leaves a very inappropriate number!! Your healthcare provider is likely to ask that you establish your baseline from which you progress slowly, perhaps increasing your time at an aerobic pace only every two weeks. This is not a race to the finish. This is about you following your healthcare provider's guidance in how far and how fast you can progress. Ideally you can work up to, or close to, 45–60 minutes per day, three to five days a week, with your heart rate in the 70–80% range as recommended by Baker[30] based on her recent and ongoing research. But you might want to start at the 50% level, depending on your current level of fitness. Follow your healthcare provider's advice.

Be metric. Begin by establishing your baseline activity:

How many minutes per day are you doing the specific exercise?
How many days per week are you currently doing the specific exercise?
What is your heart rate:

- When resting?
- During normal activity; and
- The range of your heart rate during your exercise?

Be metric across time. Measure baseline; measure each day; and CELEBRATE:

1) That you are exercising (because many people cannot).
2) Your gradual increases in intensity, frequency, and duration of exercise sessions.
3) Your gradual increases in clear thinking.
4) Improvements in your mood.
5) Increases in the speed of your thinking.
6) Increases in the speed of your actions.
7) Improvements in your memory, AND
8) Improvements in your executive skills of reasoning and planning.

Take exercise breaks on the job because even brief episodes of exercise can energize.

CELEBRATE because positive reinforcement can enhance progress!! 40 minutes to 1 hour, 3–4 days a week is a goal to work up to because aerobic exercise could increase birth, survival and health of neurons that also have an enriched environment.

Be aerobic: Several hours before going to bed to establish a sleep debt.

The earlier in the day the better since your core body temperature must return to normal to enhance sleep.

Summary:
Use them and they will grow!!
You want your brain volume to be greater than normal for your age!!
For Fun and Function!!

References

1. Flynn JR. (2012). Are We Getting Smarter? Rising IQ in the Twenty-First Century. New York: Cambridge University Press.
2. Flynn JR. (1987). Massive IQ Gains in 14 Nations: What IQ Tests Really Measure. *Psychol Bull*, 101: 171–191.
3. Flynn JR. (2018). Reflections about Intelligence over 40 Years. *Intelligence*, 70: 73–83. https://doi.org/10.1016/j.intell.2018.06.007
4. Billat V, Dhonneur Gs & Mille-Hamard, L. (2017). Case Studies in Physiology: Maximal Oxygen Consumption and Performance in a Centenarian Cyclist. *J App Physio*, 122: 430–434. https://doi.org/10.1152/japplphysiol.00569.2016
5. Sanchis-Gomar F, Pareja-Galeano, H & Lucia A. (2014). 'Olympic' Centenarians: Are They Just Biologically Exceptional? *Int J Cardiol*, 175(1). https://doi.org/10.1016/j.ijcard.2014.04.247
6. Lepers R, Stapley PJ & Cattagni T. (2016) Centenarian Athletes: Examples of Ultimate Human Performance? *Age and Ageing*, 45: 729–733. https://doi.org/10.1093/ageing/afw111
7. Bhardwaj R, Amiri S, Buchwald D & Amram O. (2020) Environmental Correlates of Reaching a Centenarian Age: Analysis of 144,665 Deaths in Washington State for 2011–2015. *Int J Environ Res Public Health*, 17: 2828. https://doi.org/10.3390/ijerph17082828
8. Zhu J, Ge F, Zeng Y, Qu Y, Chen W, Yang H, et al. (2022b) Physical and Mental Activity, Disease Susceptibility, and Risk of Dementia: A Prospective Cohort Study Based on UK Biobank. Neurology. Publish Ahead of Print. https://doi.org/10.1212/WNL.0000000000200701
9. Heo J, Noble EE & Call JA. (2023). The Role of Exerkines on Brain Mitochondria: A Mini-Review. *J Appl Physiol*, 134: 28–35. https://doi.org/10.1152/japplphysiol.00565.2022
10. Vints WAJ, Levin O, Fujiyama H, Verbunt J & Masiulis N. (2022) Exerkines and Long-Term Synaptic Potentiation: Mechanisms of Exercise-Induced Neuroplasticity. *Front Neuroendocrinol*, 66: 100993. https://doi.org/10.1016/j.yfrne.2022.100993
11. Kim E, Kim H, Jedrychowski MP, Bakiasi G, Park J, Kruskop J, et al. (2023). Irisin Reduces Amyloid-B by Inducing the Release of Neprilysin from Astrocytes Following Downregulation of ERK-STAT3 Signaling. *Neuron*, 111: 1–15. https://doi.org/10.1016/j.neuron.2023.08.012
12. Almeida OP, Khan KM, Hankey GJ, Yeap BB, Golledge J & Flicker L. (2014). 150 Minutes of Vigorous Physical Activity Per Week Predicts Survival and Successful Ageing: A Population-based 11-Year Longitudinal Study of 12,201 Older Australian Men. *Brit J Sports Medicine*, 48: 220–225.
13. Chaddock-Heyman L, Weng TB, Loui P, Weisshappel R, Drollette ES. Kienzler C, et al. (2021). Brain Network Modularity Predicts Improvements in Cognitive and Scholastic Performance in Children Involved in a Physical Activity Intervention. *Front Human Neurosc*, 14: 346.
14. Tait JL, Collyer TA, Gall SL, et al. (2022). Longitudinal Associations of Childhood Fitness and Obesity Profiles with Midlife Cognitive Function: An Australian Cohort Study. *J Sci & Med in Sport*, 25: 667–672.

15. Ward DD, Ranson JM, Wallace LMK, et al. (2022). Frailty, Lifestyle, Genetics and Dementia Risk. *J Neurol Neurosurg Psychiatry*, 93: 343–350. https://doi.org/10.1136/jnnp-2021-327396

16. Sewell KR, Rainey-Smith SR, Villemagne VL, Peiffer J, Sohrabi HR, Taddei K, et al. (2023) The Interaction Between Physical Activity and Sleep on Cognitive Function and Brain Beta-Amyloid in Older Adults. Behav Brain Res, 437: 114108.

17. Bartlett LK, Pirrone A & Javed N. (2022). Computational Scientific Discovery in Psychology. Perspec Psychol Sci. https://doi.org/10.1177/17456916221091833

18. Pereira AC, Huddleston DE & Brickman AM. (2007) An in Vivo Correlate of Exercise-Induced Neurogenesis in the Adult Dentate Gyrus. *PNAS-USA*, 104(13): 5638–5643.

19. Snyder JS, Glover LR, Sanzone KM, Kamhi JF & Cameron HA. (2009). The Effects of Exercise and Stress on the Survival and Maturation of Adult-Generated Granule Cells. *HIPPOCAMPUS*. https://doi.org/10.1002/hipo.20552

20. Angevaren M, Aufdemkampe G & Verhaar HJJ. (2008). Physical Activity and Enhanced FITNESS to Improve Cognitive Function in Older People without Known Cognitive Impairment. The Cochrane Collaboration.

21. Niemann C, Godde B & Voelcker-Rehage C. (2014) Not Only Cardiovascular, but also Coordinative Exercise Increases Hippocampal Volume in Older Adults. *Front Aging Neurosci*, 6: Article 170. https://doi.org/10.3389/fnagi.2014.00170

22. Tian Q, Erickson KI, Simonsick EM, Aizenstein H, Glynn NW, Boudreau RM, et al. (2014). Physical Activity Predicts Microstructural Integrity in Memory-Related Networks in Very Old Adults. *J Gerontol Biol Sci Med Sci*, 69(10): 1284–1290.

23. Burzynska AZ, Chaddock-Heyman L & Voss MW. (2014). Physical Activity and Cardiorespiratory Fitness Are Beneficial for White Matter in Low-Fit Older Adults. *Plos One*, 9(9): e107413.

24. Audiffren M, Tomporowski P & Zagrodnik J. (2008). Acute Aerobic Exercise and Information Processing: Energizing Motor Processes during a Choice Reaction Time Task. *Acta Psychologica*, 129: 410–419.

25. Erickson KI, Voss MW, Prakash SH, Basak C, Szabo A, et al. (2011) Exercise Training Increases Size of Hippocampus and Improves Memory. *PNAS, USA*, 108(7): 3017–3022.

26. Anzeneder S, Zehnder Z, Schmid J, Martin-Niedecken AL, Schmidt M, Benzig V, et al. (2023). Dose–Response Relation between the Duration of a Cognitively Challenging Bout of Physical Exercise and Children's Cognition. *Scand J Med Sci Sports*, 33: 1439–1451. https://doi.org/10.1111/sms.14370

27. González-Fernández FT, Delgado-Garcia G, Coll JS, Silva AF, Nobari H & Clemente FM. (2023). Relationship between Cognitive Functioning and Physical Fitness in Regard to Age and Sex. BMC Pediatr, 23(1): 204. https://doi.org/10.1186/s12887-023-04028-8

28. Bliss ES, Biki SM, Wong RHX, Howe PRC & Mills DE. (2023). The Benefits of Regular Aerobic Exercise Training on Cerebrovascular Function and Cognition in Older Adults. *Eur J Appl Physiol*, 123: 1323–1342. https://doi.org/10.1007/s00421-023-05154-y

29. Baker LD, Frank LL & Foster-Schubert K. (2010). Effects of Aerobic Exercise on Mild Cognitive Impairment: A Controlled Trial. *Arch Neurol*, 67(1): 71–79.

30. Baker LD. (2015). Aerobic Exercise Reduces CSF Levels of Phosphorylated Tau in Older Adults with MCI. For the Alzheimer's Association International Conference® 2015 (AAIC® 2015).

31. Malmo V, Nes BM, Amundsen BH, Tjonna AE, Stoylen A, Rossvoll O, et al. (2016). Aerobic Interval Training Reduces the Burden of Atrial Fibrillation in the Short Term A Randomized Trial. *Circulation*, 133: 466–473. https://doi.org/10.1161/CIRCULATIONAHA.115.018220

32. Erickson KI, Leckie RL & Weinstein AM. (2014). Physical Activity, Fitness, and Gray Matter Volume. *Neurobiol Aging*, 35: 20–28.

33. Henrique PPB, Perez FMP & Dorneles G. (2023) Exergame and/or Conventional Training-Induced Neuroplasticity and Cognitive Improvement by Engaging Epigenetic and Inflammatory Modulation in Elderly Women: A Randomized Clinical Trial. *Physiol Behav*, 258: 113996.

34. Huang B, Chen K & Li Y. (2023) Aerobic Exercise, an Effective Prevention and Treatment for Mild Cognitive Impairment. *Front Aging Neurosci*, 15: 1194559. https://doi.org/10.3389/fnagi.2023.1194559

35. Yang J, Dong Y, Yan S, Yi L & Qiu J. (2023). Which Specific Exercise Models Are Most Effective on Global Cognition in Patients with Cognitive Impairment? A Network Meta-Analysis. *Int J Environ Res Public Health*, 20: 2790. https://doi.org/10.3390/ijerph20042790

36. Fari G, Lunetti P & Pignatelli G. (2021) The Effect of Physical Exercise on Cognitive Impairment in Neurodegenerative Disease: From Pathophysiology to Clinical and Rehabilitative Aspects. *Int J Mol Sci*, 27: 22(21): 11632.

37. Xu L, Gu H, Cai X, Zhang Y, Hou X, Yu J & Sun T. (2023). The Effects of Exercise for Cognitive Function in Older Adults: A Systematic Review and Meta-Analysis of Randomized Controlled Trials. *Int J Environ Res Public Health*, 20: 1088. https://doi.org/10.3390/ijerph20021088

38. Kuo H-I, Hseih M-H, Lin Y-T, et al. (2023). A Single Bout of Aerobic Exercise Modulates Motor Learning Performance and Cortical Excitability in Humans. *Itnl J Clin Health Psychol*. https://doi.org/10.1016/j.ijchp.2022.100333

39. Won J, Nielson KA & Smith JC. (2023). Large-Scale Network Connectivity and Cognitive Function Changes After Exercise Training in Older Adults with Intact Cognition and Mild Cognitive Impairment. *J Alzheimer's Dis Reports*, 7: 399–413. https://doi.org/10.3233/ADR-220062

40. Bovolini A, Costa-Brito AR, Martins F, Furtado GE, Mendonça GV & Vila-Chã C. (2022). Impact of Exercise on Vascular Function in Middle-Aged and Older Adults: A Scoping Review. *Sports*, 10: 208. https://doi.org/10.3390/sports10120208

41. Su S, Shi L, Zheng Y, Sun Y, Huang X, Zhang A, Que J, Sun X, et al. (2022). Leisure Activities and the Risk of Dementia: A Systematic Review and Meta-Analysis. *Neurology*. https://doi.org/10.1212/WNL.0000000000200929

42. Hofman A, Rodriguez-Ayllon M, Vernooij MW, Croll PH, Luik AI, Neumann A, et al. (2023). Physical Activity Levels and Brain Structure in Middle-Aged and Older Adults: A Bidirectional Longitudinal Population-Based Study. *Neurobio of Aging*, 121: 28–37.

43. Graciani AL, Gutierre MU, Coppi AA, Arida RM & Gutierre RC. (2023). Myelin, Aging, and Physical Exercise. Neurobiol Aging, 127: 70–81. https://doi.org/10.1016/j.neurobiolaging.2023.03.009

44. Karamacoska D, Butt A, Leung IHK, Childs RL, Metri N-J, Uruthiran V, Tan T, Sabag, A & Steiner-Lim GZ. (2023) Brain Function Effects of Exercise Interventions for Cognitive Decline: A Systematic Review and Meta-Analysis. *Front Neurosci*, 17: 1127065. https://doi.org/10.3389/fnins.2023.1127065

45. Arosio B, Calvani R, Ferri E Coelho-Junior HJ, Carandina A, Campanelli F, Ghiglieri V, Marzetti E & Picca A. (2023). Sarcopenia and Cognitive Decline in Older Adults: Targeting the Muscle–Brain Axis. *Nutrients*, 15: 1853. https://doi.org/10.3390/nu15081853

46. Jemni M, Zaman R, Carrick FR, Clarke ND, Marina M, Bottoms L, Matharoo JS, et al. (2023). Exercise Improves Depression through Positive Modulation of Brain-Derived Neurotrophic Factor (BDNF). A Review Based on 100 Manuscripts over 20 Years. *Front Physiol*, 14: 1102526. https://doi.org/10.3389/fphys.2023.1102526

47. Kitamura K, Watanabe Y, Kabasawa K, Takahashi A, Saito T, Kobayashi R, et al. (2022) Leisure-Time and Non-Leisure-Time Physical Activities Are Dose-Dependently Associated with a Reduced Risk of Dementia in Community-Dwelling People Aged 40–74 Years: The Murakami Cohort Study. *JAMDA*, 23: 1197e1204 https://doi.org/10.1016/j.jamda.2022.01.053

48. Morris JK, Vidoni ED & Johnson DK. (2017) Aerobic Exercise for Alzheimer's Disease: A Randomized Controlled Pilot Trial. *PLoS ONE*, 12(2): e0170547.

49. Mitchell JJ, Blodgett JM, Chastin SFM, Jefferis BJ, Wannamethee SG & Hamer M. (2022). Exploring the Associations of Daily Movement Behaviours and Mid-Life Cognition: A Compositional Analysis of the 1970 British Cohort Study. J Epidemiol Community Health, Epub ahead of print: https://doi.org/10.1136/jech2022–219829

50. Sprague BN, Zhu X, Rosso AL, Verghese J, Delbaere KI, Lipnicki DM, Sachdev PS, et al. (2023). Correlates of Gait Speed among Older Adults from Six Countries: Findings from the COSMIC Collaboration. *J Gerontol A Biol Sci Med Sci*, Mar 28: glad090. https://doi.org/10.1093/gerona/glad090. Online ahead of print.

51. Hwang D-J, Koo J-H, Kim T-K, Jang Y-C, Hyun A-H, Yook J-S, Yoon C-S & Cho J-Y. (2023). Exercise as an Antidepressant: Exploring Its Therapeutic Potential. *Front Psychiatry*, 14: 1259711. https://doi.org/10.3389/fpsyt.2023.1259711

52. Azevedo CV, Hashiguchi D, Campos HC, Figueiredo EV, Otaviano SFSD, Penitente AR, Arida RM & Longo BM. (2023) The Effects of Resistance Exercise on Cognitive Function, Amyloidogenesis, and Neuroinflammation in Alzheimer's Disease. *Front Neurosci*, 17: 1131214. https://doi.org/10.3389/fnins.2023.1131214

53. Herold F, Törpel A, Schega L & Müller NG. (2019). Functional and/or Structural Brain Changes in Response to Resistance Exercises and Resistance Training Lead to Cognitive Improvements – A Systematic Review. *Eur Rev Aging Phys Act*, 16: 10. https://doi.org/10.1186/s11556-019-0217-2

54. Broadhouse KM, Singh MF & Suo C. (2020). Hippocampal Plasticity Underpins Long-Term Cognitive Gains from Resistance Exercise in MCI. NeuroImage: Clinical, 102182.

55. Meysami S, Raji CA, Glatt RM, Popa ES, Ganapathi AS, Bookheimer T, et al. (2023). Handgrip Strength Is Related to Hippocampal and Lobar Brain Volumes in a Cohort of Cognitively Impaired Older Adults with Confirmed Amyloid Burden. *J Alzheimer's Disease*, 91: 999–1006. https://doi.org/10.3233/JAD-220886

56. Prokopidis K, Giannos P, Ispoglou T, Kirk B, Witard C, Dionyssiotis Y, Scott D, Macpherson H, Duque G & Isanejad M. (2023). Handgrip Strength Is Associated with Learning and Verbal Fluency in Older Men without Dementia: Insights from the NHANES. *GeroSci*, 45: 1049–1058. https://doi.org/10.1007/s11357-022-00703-3

57. Cui L, Tao S, Yin H-c, Shen Q-q, Wang Y, Zhu L-n & Li X-j. (2021). Tai Chi Chuan Alters Brain Functional Network Plasticity and Promotes Cognitive Flexibility. *Front Psychol*, 12: 665419. https://doi.org/10.3389/fpsyg.2021.665419

58. Zhou M, Liao H, Sreepada LP, Ladner JR, Balschi JA & Lin AP. (2018). Tai Chi Improves Brain Metabolism and Muscle Energetics in Older Adults. *J Neuroimaging*, 28: 359–364. https://doi.org/10.1111/jon.12515

59. Yue C, Yu Q, Zhang Y, Herold F, Mei J, Kong Z, Perrey S, Liu J, Müller NG, Zhang Z, Tao Y, Kramer A, Becker B & Zou L. (2020). Regular Tai Chi Practice Is Associated with Improved Memory as Well as Structural and Functional Alterations of the Hippocampus in the Elderly. *Front Aging Neurosci*, 12: 586770. https://doi.org/10.3389/fnagi.2020.586770

60. Yang Y, Chen T, Wang C, Zhang J, Yuan X, Zhong X, Yan S & Jiang C. (2022). Determining Whether Tai Chi Chuan Is Related to the Updating Function in Older Adults: Differences Between Practitioners and Controls. *Front Public Health*, 10: 797351. https://doi.org/10.3389/fpubh.2022.797351

61. Chen W, Zhang X, Sie H, He Q & Shi Z. (2022). Brain Functional Connectivity in Middle-Aged Hong Chuan Tai Chi Players in Resting State. *Int J Environ Res Public Health*, 19: 12232. https://doi.org/10.3390/ijerph191912232

62. Siu M-Y & Lee DTF (2018). Effects of Tai Chi on Cognition and Instrumental Activities of Daily Living in Community Dwelling Older People with Mild Cognitive Impairment. *BMC Geriatrics*, 18: 37. https://doi.org/10.1186/s12877-018-0720-8

63. Siu M-Y & Lee DTF (2021). Is Tai Chi an Effective Intervention for Enhancing Health-Related Quality of Life in Older People with Mild Cognitive Impairment? An Interventional Study. *Int J Older People Nurs*, 16: e12400. https://doi.org/10.1111/opn.12400

64. Li F, Harmer P, Fitzgerald K & Winters-Stone K. (2022). A Cognitively Enhanced Online Tai Ji Quan Training Intervention for Community-Dwelling Older Adults with Mild Cognitive Impairment: A Feasibility Trial. *BMC Geriatrics*, 22: 76. https://doi.org/10.1186/s12877-021-02747-0

65. Wang C, Dai Y, Yang Y, Yuan X, Zhang M, Zeng J, Zhong X, Meng J & Jiang C. (2023). Effects of Tai Chi on Working Memory in Older Adults: Evidence from Combined fNIRS and ERP. *Front Aging Neurosci*, 15: 1206891. https://doi.org/10.3389/fnagi.2023.1206891

66. Kilpatrick LA, Siddarth P, Milillo MM, Sorio BK, Ercoli L, Narr KL & Lavretsky H. (2022). Impact of Tai Chi as an Adjunct Treatment on Brain Connectivity in Geriatric Depression. *J Affect Disord*, 315: 1–6. https://doi.org/10.1016/j.jad.2022.07.049. Epub 2022 July 26.

67. Jeong G, Jung H, So W-Y & Chun B. (2023). Effects of Taekwondo Training on Growth Factors in Normal Korean Children and Adolescents: A Systematic Review and Meta-Analysis of Randomized Controlled Trials. *Children*, 10: 326. https://doi.org/10.3390/children10020326
68. Kim YJ, Cha EJ, Kim SM, Kang KD & Han DH. (2015). The Effects of Taekwondo Training on Brain Connectivity and Body Intelligence. *Psychiatry Investig*, 12(3): 335–340. https://dx.doi.org/10.4306/pi.2015.12.3.335
69. Kurtoğlu E, Payas A, Düz S, Arik M, Uçar I, Tokmak TT, et al. (2023). Analysis of Changes in Brain Morphological Structure of Taekwondo Athletes by Diffusion Tensor Imaging. *J Chem Neuroanat*, 129: 102250. https://doi.org/10.1016/j.jchemneu.2023.102250
70. Cho SY, Kim YI & Roh HT. (2017a). Effects of Taekwondo Intervention on Cognitive Function and Academic Self-Efficacy in Children. *J Phys Ther Sci*, 29(4): 713–715. https://doi.org/10.1589/jpts.29.713. Epub 2017 Apr 20.
71. Cho SY, So WY & Roh HT. (2017b). The Effects of Taekwondo Training on Peripheral Neuroplasticity-Related Growth Factors, Cerebral Blood Flow Velocity, and Cognitive Functions in Healthy Children: A Randomized Controlled Trial. *Int J Environ Res Public Health*, 14(5): 454. https://doi.org/10.3390/ijerph14050454.
72. Cho SY & Roh HT. (2019). Taekwondo Enhances Cognitive Function as a Result of Increased Neurotrophic Growth Factors in Elderly Women. *Int J Environ Res Public Health*, 16(6): 962. https://doi.org/10.3390/ijerph16060962.
73. Kadri A, Slimani M, Bragazzi NL, Tod D & Azaiez F. (2019). Effect of Taekwondo Practice on Cognitive Function in Adolescents with Attention Deficit Hyperactivity Disorder. *Int J Environ Res Public Health*, 16: 204. https://doi.org/10.3390/ijerph16020204
74. Van Dijk GP, Huijts M & Lodder J. (2013). Cognition Improvement in Taekwondo Novices Over 40. Results from the SEKWONDO Study. *Front Aging Neurosci*, 5: 74. https://doi.org/10.3389/fnagi.2013.00074.
75. Andrade-Lara KE, Latorre Román PÁ; Párraga Montilla JA & Cabrera Linares JC. (2023). Can 16 Minutes of HIIT Improve Attentional Resources in Young Students? *J Funct Morphol Kinesiol*, 8: 116. https://doi.org/10.3390/jfmk8030116
76. Martínez-Díaz IC & Carrasco Páez L. (2023) Little but Intense: Using a HIIT-Based Strategy to Improve Mood and Cognitive Functioning in College Students. *Healthcare*, 11: 1880. https://doi.org/10.3390/healthcare11131880
77. Guo L, Chen J & Yuan W (2023) The Effect of HIIT on Body Composition, Cardiovascular Fitness, Psychological Wellbeing, and Executive Function of Overweight/Obese Female Young Adults. *Front Psychol*, 13: 1095328. https://doi.org/10.3389/fpsyg.2022.1095328
78. Molinaro L, Taborri J, Pauletto D, Guerra V, Molinaro D, Sicari G, et al. (2023). Measuring the Immediate Effects of High-Intensity Functional Training on Motor, Cognitive and Physiological Parameters in Well-Trained Adults. Sensors (Basel), 23(8): 3937. https://doi.org/10.3390/s23083937
79. Norling AM, Gerstenecker A, Bolding MS, Ver Hoer L, Buford T, Walden R, An H, et al. (2023). Effects of a Brief HIIT Intervention on Cognitive Performance in Older Women. *Geroscience*, Aug 15. https://doi.org/10.1007/s11357-023-00893-4. Online ahead of print.

80. Clemente-Suárez VJ, Villafaina S, García-Calvo T & Fuentes-García JP. (2022). Impact of HIIT Sessions with and without Cognitive Load on Cortical Arousal, Accuracy and Perceived Exertion in Amateur Tennis Players. *Healthcare*, 10: 767. https://doi.org/10.3390/healthcare10050767

81. Oliva HNP, Oliveira GM, Oliva IO, Cassilhas RC, Batista de Paula AM & Monteiro-Junior RS. (2023). Middle Cerebral Artery Blood Velocity and Cognitive Function after high- and Moderate-Intensity Aerobic Exercise Sessions. *Neurosci Lett*, 817: 137511. https://doi.org/10.1016/j.neulet.2023.137511. Online ahead of print.

82. Lee S, Choi Y, Jeong E, Park J, Kim J, Tanaka M & Choi J. (2022) Physiological Significance of Elevated Levels of Lactate by Exercise Training in the Brain and Body. *J Bioscience and Bioengineering*, 135(3): 167e175. https://doi.org/10.1016/j.jbiosc.2022.12.001

83. Kempermann G, Song H & Gage FH. (2023). Adult Neurogenesis in the Hippocampus. *Hippocampus*, 33(4): 269–270. https://doi.org/10.1002/hipo.23525

84. Choi SH & Tanzi RE. (2023) Adult Neurogenesis in Alzheimer's Disease. *Hippocampus*, 33: 307–321. https://doi.org/10.1002/hipo.23504

85. MusicMendsMinds.org.

86. Shaffer J. (2016). Neuroplasticity and Clinical Practice: Building Brain Power for Health. Front Psychol, 7: 1118. https://doi.org/10.3389/fpsyg.2016.01118

87. PlayingForChange.com.

88. Carapellotti AM, Meijerink HJEM, Gravemaker-Scott V, Thielman L, Kool R, et al. (2023). Escape, Expand, Embrace: The Transformational Lived Experience of Rediscovering the Self and the other While Dancing with Parkinson's or Multiple Sclerosis. *Intnl J Qual Studies Health Well-Being*, 18: 2143611. https://doi.org/10.1080/17482631.2022.2143611

89. Meulenberg CJW, Rehfeld K, Jovanović S, & Marusic U. (2023). Unleashing the Potential of Dance: A Neuroplasticity-Based Approach Bridging from Older Adults to Parkinson's Disease Patients. *Front Aging Neurosci*, 15: 1188855. https://doi.org/10.3389/fnagi.2023.1188855

90. Esteban-Cornejo I, Rodriguez-Ayllon M, Verdejo-Roman J, Cadenas-Sanchez C, Mora-Gonzalez J, Chaddock-Heyman L, Raine LB, et al. (2019). Physical Fitness, White Matter Volume and Academic Performance in Children: Findings From the ActiveBrains and FITKids2 Projects. *Front Psychol*, 10: 208. https://doi.org/10.3389/fpsyg.2019.00208

91. Muinos J & Ballesteros S (2021) Does Dance Counteract Age-Related Cognitive and Brain Declines in Middle-Aged and Older Adults? A Systematic Review. *Neurosc Biobeh Rev*, 121: 259–276.

92. Burzynska AZ, Finc K, Taylor BK, Knecht AM & Kramer AF. (2017). The Dancing Brain: Structural and Functional Signatures of Expert Dance Training. *Front Hum Neurosci*, 11: 566. https://doi.org/10.3389/fnhum.2017.00566

93. Balazova Z, Marecek R, Novakova L, Nemcova-Elfmarkova N, Kropacova S, Brabenec L, Grmela R, Vaculíková P, Svobodova L & Rektorova I. (2021). Dance Intervention Impact on Brain Plasticity: A Randomized 6-Month fMRI Study in Non-expert Older Adults. *Front Aging Neurosci*, 13: 724064. https://doi.org/10.3389/fnagi.2021.724064

94. Podolski OS, Whitfield T, Schaaf L, Cornaro C, Köbe T, Koch S & Wirth M. (2023). The Impact of Dance Movement Interventions on Psychological Health in Older Adults without Dementia: A Systematic Review and Meta-Analysis. *Brain Sci*, 13: 981. https://doi.org/10.3390/brainsci13070981

95. Wu VX, Chi Y, Le JK, Goh HS, Chen DTM, Haugen G, Chao FFT, et al. (2021). The Effect of Dance Interventions on Cognition, Neuroplasticity, Physical Function, Depression, and Quality of Life for Older Adults with Mild Cognitive Impairment: A Systematic Review and Meta-Analysis. *Int J Nur St*, 122: 104025.

96. Zhu Y, Gao Y, Guo C, Qi M, Xiao M, Wu H, Ma J, et al. (2022a). Effect of 3-Month Aerobic Dance on Hippocampal Volume and Cognition in Elderly People With Amnestic Mild Cognitive Impairment: A Randomized Controlled Trial. *Front Aging Neurosci*, 14: 771413. https://doi.org/10.3389/fnagi.2022.771413

97. Wang L, Li F & Tang L (2022a). Chronic Effects of Different Exercise Types on Brain Activity in Healthy Older Adults and those with Parkinson's Disease: A Systematic Review. *Front Physiol*, 13: 1031803. https://doi.org/10.3389/fphys.2022.1031803

98. Wu X, Lu X, Zhang H, Wang X, Kong Y & Hu L. (2023) The Association between Ballroom Dance Training and Empathic Concern: Behavioral and Brain Evidence. *Hum Brain Mapp*, 44: 315–326. https://doi.org/10.1002/hbm.26042

99. Clifford AM, Shanahan J, McKee J, Cleary T, O'Neill A, O'Gorman M, Louw Q & Bhriain ON. (2023). The Effect of Dance on Physical Health and Cognition in Community Dwelling Older Adults: A Systematic Review and Meta-Analysis. *Arts Health*, 15(2): 200–228. https://doi.org/10.1080/17533015.2022.2093929.

100. Sato K, Ochi A, Watanabe K & Yamada K. (2023). Effects of Dance Video Game Training on Cognitive Functions of Community-Dwelling Older Adults with Mild Cognitive Impairment. *Aging Clin Exp Res*, 35(5): 987–994. https://doi.org/10.1007/s40520-023-02374-2. Epub 2023 Mar 3.

101. Rodrigues-Krause J, Krause M & Reischak-Oliveira A. (2019). Dancing for Healthy Aging: Functional and Metabolic Perspectives. *Altern Ther Health Med*, 25(1): 44–63.

102. Nascimento MDM. (2021). Dance, Aging, and Neuroplasticity: An Integrative Review. Neurocase. https://doi.org/10.1080/1354.2021.1966047.

103. Raghupathy MK, Divya M & Karthikbabu S. (2022). Effects of Traditional Indian Dance on Motor Skills and Balance in Children with Down syndrome. *J Mot Behav*, 54(2): 212–221. https://doi.org/10.1080/00222895.2021.1941736. Epub 2021 July 8.

104. Kosurko A, Herron RV, Grigorovich A, et al. (2022). Dance Wherever You Are: The Evolution of Multimodal Delivery for Social Inclusion of Rural Older Adults. Innov Aging, 6(2): 1–12.

105. Chan JSY, Wu J, Deng K & Yan JH (2020). The Effectiveness of Dance Interventions on Cognition in Patients with Mild Cognitive Impairment: A Meta-Analysis of Randomized Controlled Trials. *Neurosci Biobehav Rev*, 118: 80–88. https://doi.org/10.1016/j.neubiorev.2020.07.017

106. Wang Y, Liu M, Tan Y, Dong Z, Wu J, Cui H, et al (2022b). Effectiveness of Dance-Based Interventions on Depression for Persons with MCI and Dementia: A Systematic Review and Meta-Analysis. *Front Psychol*, 12: 709208.

107. Won J, Nielson KA & Smith JC. (2023). Large-Scale Network Connectivity and Cognitive Function Changes After Exercise Training in Older Adults with Intact Cognition and Mild Cognitive Impairment. *J Alzheimer's Disease Rep*, 7: 399–413. https://doi.org/10.3233/ADR-220062

108. Camuso, S & Canterini, S. (2023). Brain-Derived Neurotrophic Factor in Main Neurodegenerative Diseases. Neural Regen Res, 18(3): 554–555.

109. Erickson KI, Prakash RS, Voss MW, Chaddock L, Heo S, et al. (2010). Brain-Derived Neurotrophic Factor Is Associated with Age-Related Decline in Hippocampal Volume. *J Neurosc*, 30(15): 5368–5375.

110. Lawry T. (2023). Hacking Health Care: How AI and the Intelligence Revolution Will Reboot an Ailing System. New York: Routledge.

111. Maugeri GV & Musumeci G. (2023). Role of Exercise in the Brain: Focus on Oligodendrocytes and Remyelination. Neural Regen Res, 18(12): 2645–2646.

112. Diamond, MC. (1988). *Enriching Heredity: The Impact of the Environment on the Anatomy of the Brain*. New York: The Free Press.

113. Colcombe S & Kramer AF. (2003) Fitness Effects on the Cognitive Function of Older Adults: A Meta-Analytic Study. *Psychological Sci*, 14(2).

114. Islam MR, Valaris S, Young MF, Haley EB, Luo R, Bond SF, et al (2021). Exercise Hormone Irisin Is a Critical Regulator of Cognitive Function. *Nat Metab*, 3(8): 1058–1070. https://doi.org/10.1038/s42255-021-00438-z

115. Gage F. (2002). Neurogenesis in the Adult Brain. *J11 Neuroscience*, 22(3): 612–613.

116. Feng L, Yin J, Zhang P, An J, Zhao Y, Song Q, Ping P, et al. (2023). Physical Inability Rather than Depression and Cognitive Impairment Had Negative Effect on Centenarian Prognosis: A Prospective Study with 5-Year Follow-Up. *J Affect Disord*, 338: 299–304. https://doi.org/10.1016/j.jad.2023.05.072. Epub 2023 May 25.

117. Jessberger S. & Gage FH. (2008). Stem-Cell-Associated Structural and Functional Plasticity in the Aging Hippocampus. *Psychology & Aging*, 23(4): 684–691.

118. Kweon J, Vigne MM, Jones RN, Carpenter LL & Brown JC (2023). Practice Makes Plasticity: 10-Hz rTMS Enhances LTP-Like Plasticity in Musicians and Athletes. *Front Neural Circuits*, 17: 1124221. https://doi.org/10.3389/fncir.2023.1124221

119. Mou H, Fang Q, Tian S & Qiu F. (2023). Effects of Acute Exercise with Different Modalities on Working Memory in Men with High and Low Aerobic Fitness. *Physio Behav*, 258: 114012.

120. Hortobagyi T, Vetrovsky T, Balbim GM, et al. (2022). The Impact of Aerobic and Resistance Training Intensity on Markers of Neuroplasticity in Health and Disease. *Ageing Res Rev*, 80: 101698.

121. Kundu S, Huang H, Erickson KI, et al. (2021). Investigating Impact of Cardiorespiratory Fitness in Reducing Brain Tissue Loss Caused by Ageing. *Brain Commun*, 3(4): https://doi.org/10.1093/braincomms/fcab228

122. Logan NE, Henry DA, Hillman CH & Kramer AF. (2022): Trained Athletes and Cognitive Function: A Systematic Review and Meta-Analysis. *Internl J Sport Exercise Psychol*. https://doi.org/10.1080/1612197X.2022.2084764

123. Ni C, Ji Y, Hu K, Xing K, Xu Y & Gao Y. (2023). Effect of Exercise and Antioxidant Supplementation on Cellular Lipid Peroxidation in Elderly Individuals: Systematic Review and Network Meta-Analysis. *Front Physiol*, 14: 1113270. https://doi.org/10.3389/fphys.2023.1113270

124. Song Q, Cheng X & Zheng R. (2022). Effects of Different Exercise Intensities of Race-Walking on Brain Functional Connectivity as Assessed by Functional Near-Infrared Spectroscopy. *Front Hum Neurosci*, 16: 1002793. https://doi.org/10.3389/fnhum.2022.1002793

125. Tinney E, Loui P, Raine, L, Hiscox LV, Delgorio PL, Kramer MK, et al. (2022). Influence of Mild Cognitive Impairment and Body Mass Index on White Matter Integrity Assessed by Diffusion Tensor Imaging. Authorea. https://doi.org/10.22541/au.167179418.88228052/v1

126. Raji CA, Merrill DA, Eyre H, Mallam S, Torosyan N, Erickson KI, Lopez OL, et al. (2016). Longitudinal Relationships between Caloric Expenditure and Gray Matter in the Cardiovascular Health Study. *J Alzheimers Dis*, 52(2): 719–729. https://doi.org/10.3233/JAD-160057

127. Yu J, Collinson SL, Liew TM, Ng TP, Mahendren R, Kua EH & Feng L. (2020). Super-Cognition in Aging: Cognitive Profiles & Associated Lifestyle Factors. *Appl Neuropsych*, 27(6): 497–503.

128. Ciria LF, Roman-Caballero R, Vadillo MA, Holgado D, Luque-Casado A, Perakakis P & Sanabria D. (2023). An Umbrella Review of Randomized Control Trials on the Effects of Physical Exercise on Cognition. *Nat Hum Behav*, 7: 928–941. https://doi.org/10.1038/s41562-023-01554-4

6

Neurobics

Newness and Challenge for Neurons

Neurobics is a term coined by Lawrence Katz[1] describing exercises using all five senses in ways that are different than usual, fun, portable, and can be done anytime to stimulate brain cells and the nutrients needed to retain

Figure 6.1
This illustration by David Yu, MD, of a brain cell on multiple computers captures complex new learning in one of the ways many busy people today pursue brain stimulation for growth.

DOI: 10.4324/9781003462354-6

brain health and strength. Figure 6.1 illustrates that concept. Neurobics are essential exercises for your neurons. Katz recommended neurobics for preventing loss as well as for building brain fitness.

Complex new learning is the essence of neurobics. Complex enough to hold your attention and concentration but not so complex that it is discouraging. New means it is something different within your area of expertise as well as new skills and topics which are completely outside of your realm of expertise and customary activities. An essential component of neurobics is that you are learning something that is new and at least slightly more complex than you knew before.

Each of us will have our own unique neurobics. Mine include episodic literature searches for new evidence-based neurocognitive interventions. Other favorites are computerized cognitive training programs that reinforce speed of processing and visual-motor coordination. Taking music lessons was part of my youth and is on my list of future neurobics.

> "Just as ideal forms of physical exercise emphasize using many *different muscle groups* to enhance coordination and flexibility, the ideal brain exercises involve activating many *different brain areas* in novel ways to increase the range of mental motion."
>
> *Katz and Rubin[1] in Keep Your Brain Alive*

In his use of the term neurobics, Katz encourages activities that combine multiple senses in novel ways. To the usual list of senses – sight, sound, touch, taste, and smell – he added emotions as a sixth sense. That is a wise addition because memory can be markedly improved when there are strong emotions involved; of note, memory can also be significantly impaired during traumatic events.

Katz believed that episodically changing your path to work and to home as well as modifying your ways of doing things would forge new connections in your brain that would serve you well across time. For example, he suggested that you might change to brushing your teeth with your left hand if you have always brushed with your right hand. Novelty, focused attention, challenge, and learning are essential to improving how your brain performs and grows. This is true *at any age*.

The neurobics focus here will be on computerized cognitive training, music, dance, being bilingual, and creativity. Each of these are important neurobics in their own way, just as any career-focused neurobics will be unique to the individual seeking to drive their neuroplasticity in a positive direction.

It used to be essential for people to invent neurobics in every possible situation and time. That is less necessary in the age of technology. Neurobics come to most of us easily and often with such things as hardware and software updates. Using this wisely can yield impressive brain benefits.

Animal and human research has shown that environmental stimulation is critical for enhancing and maintaining cognitive function. Novelty, focused attention, and challenge are essential components of enhancing cognitive function.[2,3] Perceived challenge is associated with enjoyment of the task[4]; it functions as positive reinforcement for humans to help maintain time on task.

Driving neuroplasticity in a positive direction with interventions associated with improved cognitive functioning and other health benefits could increase healthy brain aging from the first moments of life through end of life. It is urgent to include the entire lifespan in considering theoretical approaches and research data that could apply in business, career, clinical setting, and in the global community to enhance neuroplasticity at all ages for everyone. This is essential to improve business performance, productivity, health, wellness, and global statistics on neurocognitive functioning.

Although it will always be necessary across the entire lifespan to use each of the many categories of interventions that can drive neuroplasticity in a positive direction,[5] we can use artificial intelligence to personalize neurobics that could influence increasing human intelligence and enhancing brain functioning for personal, career, and business advantages. Invaluable neurobics include but are not limited to computerized cognitive training, pursuing education, learning more than one language, using a computer, accessing the internet, and creating works of art.

Computerized Cognitive Training

My appreciation of and passion for the value of computerized cognitive training was fueled by witnessing the results of one individual I worked with. Since this is a case study, it may not seem as valuable to others. However, after 15 years of math and reading remediation in special education plus some similar efforts in community college, there was no appreciable improvement in math nor in reading. However, computers can reinforce speed and accuracy with more efficiency than any human can. After this individual used a computerized program in math sufficiently, their mental math skills exceeded individuals with advanced degrees who specialized in the topic being considered. Subsequent years using their computer also built reading skills. These observations still motivate my literature searches for evidence-based computerized cognitive training as my favorite neurobic to provide newness and challenge for my neurons.

Research with the Advanced Cognitive Training for Independent and Vital Elderly (ACTIVE) random controlled trial continues to provide information on these individuals who were recruited between the ages of 65 and 95 years old.[6] With 2,832 participants, it's the largest trial to date of individuals who "did not have significant cognitive, physical, or functional decline." The trial was "designed to examine the effects of perceptual and cognitive interventions on the primary outcome of cognitively demanding tasks of daily living and secondary outcomes of health-related quality of life, mobility, and health-services utilization." The control group had no contact. There were ten sessions for each of three treatment groups: reasoning training "(ability to solve problems that follow a serial pattern)"; memory training (verbal episodic memory and they were taught memory strategies); and speed of processing which was a computer-based program requiring skills of visual searching, rapid identification of visual information, visual identification, maintaining attention, and location of stimuli.

Karlene Ball and colleagues in the ACTIVE study[7] found that as little as ten 60–75-minute sessions over a 5- to 6-week period of cognitive training helped normal elderly humans improve function on the specific skills of training with effect sizes that were "comparable with or greater than the amount of longitudinal decline that has been reported in previous studies"; they opined that this suggests "these interventions have the potential to reverse age-related decline." Computerized speed of processing training showed immediate gains regardless of age, gender, mental status, health status or education with gains maintained over five years.[8]

Participants having four, one-hour booster sessions at 11 and 35 months had gains beyond those found at immediate measures and "counteracted nearly 5 months of normative age-related decline in processing speed." With completion of all training and booster sessions, these people demonstrated about 2.5 standard deviations of *gain* in their speed of processing.

Ten years after their initial training in the ACTIVE study, gains were evident in the targeted cognitive skills of reasoning and speed of processing but not for memory. All participants assessed at ten years reported less difficulty with instrumental activities of daily living.[9] This randomized, controlled, single blind trial involving 2,802 humans in six cities in the USA is the largest trial to date.

A review with a meta-analysis of research on Useful Field of View (UFOV) training which was part of the ACTIVE study considered 44 studies from 17 random controlled trials with adults.[10] Enhanced neural outcomes included improved attention and faster speed of processing. These research reviewers opined that, since it also enhanced quality of life, health, and well-being across time, use of UFOV training for elders could improve their "real-world functional outcomes and well-being."

Of individuals in this ACTIVE research group, some were considered to be at high risk for future falling.[11] "However, the participants at greater risk for future falls in the speed of processing training group were 31% less likely" to fall again "across ten years compared to the control group." These researchers opined that although "mobility is largely physical, walking safely in one's environment necessitates maintaining speed while simultaneously monitoring for intrinsic (e.g., balance) and extrinsic (e.g., tripping hazards such as a rug) falling factors and are cognitive tasks."

Computerized cognitive training has been shown to have a powerful influence on neurocognitive skills. Working with adults aged 60–89 using a computer training program, the duration of and presentation speed of stimuli adapted to the skill level of the person to maintain a degree of challenge.[12] As a result of practice and challenge, individuals had improved working memory for tasks they had not been trained in. Pre-training and post-training electroencephalography also showed functional brain plasticity. The sample size was small ($N = 32$) and well educated (13–21 years of education).

Further research is indicated on how to maximize transfer of gains from training programs that are computer based, provide adaptive challenge, and can reinforce progress with greater accuracy and effectiveness than could ever be managed by human trainers. A recent small study with people 65 and older who were cognitively intact found that less daily computer use was associated with a smaller percentage of hippocampal volume.[13] Pairing the use of these training programs with aerobic exercise could increase brain gains.[14]

Artificial intelligence with human creative reasoning could influence future studies that include neuroimaging to increase our database on evidence-based ways that computer-based interventions to increase speed of processing can help improve many factors in functioning across the lifespan. The findings of Ball and associates that some gains persisted up to ten years highlight the need for research on dose, timing and how broadly these gains might influence other factors such as memory, judgment, and prosocial behaviors.

Extensive works by Karlene Ball, Michael Merzenich, and colleagues have used behavioral techniques in computer software training that could adapt to the current level of functioning such that positive reinforcement was predominant. Merzenich has repeatedly shown that brain plasticity can be influenced in positive ways that can enrich human experience.[2,15] A review of research on elders with mild cognitive impairment found evidence of improved cognitive functioning with computerized cognitive training.[16]

Both Ball and Merzenich have found significant gains in skills that were trained. It is especially significant that these enduring positive effects of neuroplasticity-based computerized cognitive training have been found to impact such important issues as quality of life,[17] driving mobility,[18] independence,[6] improved memory and attention in humans aged 65 and older,[19,20] reduced risk of increased depression one and five years after baseline training,[21] and reduced risk of falling.[11]

Sufficient research reports exist to indicate the value of computer-based techniques to drive neuroplasticity in a positive direction; this has influenced some to "argue that neuroplasticity-based treatments will be an important part of future best-treatment practices in neurological and psychiatric medicine."[22]

Clearly, my observations in the case study described above support advocating for neuroplasticity-based interventions as an essential part of "best-treatment practices in neurological and psychiatric medicine" going forward. Time on the computer-based math program accomplished what 15 years of remediation could not. That individual's math skills improved so much that, soon after using the computerized math training program, their speed and accuracy of mental math exceeded that of people with advanced degrees. It is also noteworthy that, for many decades they continued to exceed all others in speed and accuracy of mental math. Despite this being a case study, it warrants being considered in the context of the findings of the ACTIVE study of 2,802 individuals where gains were evident with as little as ten 60–75-minute sessions of computerized cognitive training over a 5–6-week period *AND* with gains still evident ten years later.[9]

Since computerized complex new learning coupled with aerobic exercise has shown increased gains over either intervention alone in early research,[14] ballroom dancing could be another research avenue for measuring cognitive gains in an activity that combines at least physical and cognitive elements; add social aspects when dancing with others. A recent review[23] suggests that combining cognitive training with aerobic exercise in natural activities like dancing might be the most beneficial because it is multilevel with both physical and cognitive coordination required.

Since aerobic exercise is associated with increased neurogenesis in humans, and since integration of these new brain cells requires newness and challenge, research has shown increased survival rates of new brain cells in animals afforded an enriched environment; type of stimulation during aerobic exercise can influence the site of integration as well as improved cognition related to those efforts; in essence, the timing is right for increased research to develop computerized interventions that can be used in association with aerobic activity, whether on the dance floor or cardio machine, to assess potential academic, cognitive, prosocial and business benefits across a healthier longevity. Artificial intelligence could be an essential part of measuring, refining, personalizing, and providing positive reinforcement for maximizing benefits from these evidence-based interventions.

To what extent is neurogenesis relevant to information processing? New brain cells in the dentate gyrus are essential for discriminating fine differences in experiences and sensory inputs.[24] It is this awareness of differences in small details that picks up newness. It is during their sixth to eighth week that new neurons are more excitable than mature brain cells[25] permitting the plastic response of stimulated adult-born new brain cells while preserving existing neuronal function. This important period is when the per cent of new brain cells that survive and get integrated can be influenced by intrinsic and extrinsic factors.[26] It is also when less strong exciting currents are essential to cause a plastic response such as survival, integration, memory, or long-term potentiation. These new brain cells are also thought to hold associations between time-related experiences that may not be related in content like those that occur during memory flashbacks.

Even when additional stimulation was not provided until rats were middle-aged, an enriched environment resulted in a fivefold increase in neuronal phenotypes that was associated with "significant improvements in learning parameters, exploratory behavior and locomotor activity."[27] In addition, these new and enriching experiences resulted in decreased age dependent degeneration as shown by less accumulation of lipofuscin in the dentate gyrus. These findings corroborate those of Diamond's work that neuroplastic gains with new and enriched environments are not limited to a brief "critical" period.[28–30]

In their first few days of life new brain cells grow dendrites that reach into the dentate gyrus and axons are evident. It is during this young and excitable period that new neurons require neurobics and experiences such as physical activity and learning that includes challenge and newness to become integrated and stable in the dentate gyrus. Although the many steps between birth and the connectivity that includes synaptic integration are still awaiting definition, there is largely no difference between mature brain cells and newborn ones by the time new neurons are about eight weeks old.[31]

While many of the steps and mechanisms of such rapid maturity are still being studied,[24] "current hypotheses suggest that it will depend on both intrinsic signaling pathways and extrinsic regulators and local network activity." That's why you want copious neurobics for your neurons. As described above, computerized cognitive training in a skill you want to learn might be high on your list.

Knowledge alone is rarely sufficient for achieving desired outcomes. It was observing the significant gains of the individual described above with computerized cognitive training in math that initiated my interest in this type of evidence-based interventions.

People want to do what is fun, easy, portable, adapted to their culture and preferences, and at no or very little cost. Computer-based training can meet these needs while providing excellent positive influence on neuroplasticity. Using artificial intelligence to refine, personalize, remind, and provide positive reinforcements for activities associated with driving neuroplasticity in a positive direction can increase use of these activities for the health of your neurons.

Music and Dance

Of note, music and dance can also meet these needs and can be used in commercial, residential, and community settings including isolated rural communities. Music and dance can be provided in person or virtually. Not only does evolving neuroscience of music provide increasing documentation of neuroplastic purchases of music in healthy people as well as those with compromised brain structure and function; it also finds that music activities are often combined with other lifestyle choices that have been found to improve general health as well as prosocial behaviors across the lifespan.[5] With studies also documenting increases in prosocial behaviors associated with activities across the lifespan which include music and dance, perhaps music and dance can help our global society move more easily in the direction of peaceful coexistence, another potentially positive purchase for governments, for businesses, for clinical care, and for everyone.

Although it's far beyond the scope of this writing to include all available research in this vast database on music and dance, this descriptive overview considers only research reports which met rigorous scientific standards, were broad review articles or random controlled trials, and were included without a specified time span. Receptive music, making music, and dance can be considered as separate categories of influence. This has been observed across the lifespan; simply hearing music has influenced brain activity as early as in utero and has done so across the lifespan to end of life even with individuals suffering severe dementia. To emphasize that it is essential to provide brain healthy stimulation across the entire age span, this begins with research on receptive music, simply hearing music.

Listening to Music

Event-related potential (ERP) strength at birth and at 4 months, which were correlated with extent of prenatal music exposure, indicated sufficient neuronal response to sustain fetal memory into early infancy.[32] A review of research on fetal learning[33] found that stimulation with speech and music in utero "can form stimulus specific memory traces during the fetal period and effect neonatal neural system."

Listening to and auditory processing of music involve bilateral and large-scale brain networks, noted on fMRI and PET studies, for encoding, analyzing, appreciating emotional impact and for higher level cognitive functions including attention and memory.[34] Brain networks involved with emotions and reward are more highly activated by familiar than by unfamiliar music.[35] The sound of music activates mirror neurons and the core empathy network with familiar music associated with the strongest response. This fMRI study "confirms and extends empirical claims that music cognition is inextricably linked to social cognition" and could clarify ways that music "can function as a virtual social agent."

Hearing music on the farm was a constant. It was likely there when I was in utero.

Simply listening to music has shown positive purchases in the Neonatal Intensive Care Unit (NICU) where preterm infants experience extreme stress.[36] Alleviating stress could protect, perhaps even enhance, neurodevelopment during these early days of life.[37,38] Listening to music had some success in reducing this stress[39,40] and might improve memory.[41] A review of studies in five countries[42] found that hearing recorded or live music could improve the heart rate of preterm infants in NICU, stabilize their respiratory rate, decrease stress, and improve oral intake.

Forty premature infants who heard calming background melodies through headphones for eight minutes twice a day had an MRI study immediately before they heard the music the first time.[43] Another MRI was done prior to leaving NICU. These 40 premature infants in NICU, after weeks of hearing calming melodies, had

> longitudinal increase of fiber cross-section (FC) and fiber density (FD) in all major cerebral white matter fibers. Regarding cortical grey matter, FD decreased while FC and orientation dispersion index (ODI) increased, reflecting intracortical multidirectional complexification and intracortical myelination. The music intervention resulted in a significantly higher longitudinal increase of FC and ODI in cortical paralimbic regions, namely the insulo-orbito-temporopolar complex, precuneus/posterior cingulate gyrus, as well as the auditory association cortex.

Thus, premature infant brain structural changes were documented longitudinally on a micro and macro level in cortical grey matter and white brain matter. Hearing calming background melodies was associated with "intracortical complexity in regions important for socio-emotional development, known to be impaired in preterm infants." This is MRI-documented complexity of the benefits for very premature babies in NICU of simply hearing music for eight minutes twice a day for several weeks.

Neonatal ICU advances in care and technology has resulted in more premature babies surviving.[44] Longitudinal studies are needed to assess long-term neurodevelopmental effects and to guide clinical treatments in utero and thereafter[40] on the types and duration of musical interventions. Early intervention is essential to protect the neurodevelopment of these neonates and avoid neurodevelopmental disabilities such as visual impairment, cerebral palsy, delayed socialization, etc. Stabilizing vital signs as well as improving later life auditory performance can be influenced by stimulating the neonate auditory system in the NICU. Autonomic and physiological stability was achieved by maternal sound and was improved with lullabies as music therapy. "Maternal singing during kangaroo care may be recommended for providing physiological stability." Kangaroo care includes having the baby's skin touching the parent's skin.

Their review of peer-reviewed research on live music therapy potential impacts for the health and well-being of care providers, children, and their families while in pediatric hospitals helped identify the prerequisites for these interventions to provide benefit of less time in hospital, better coping, and more positive affect.[45] Play and familiarity are important aspects of music communication.

Shrock[46] is an advocate for daily music to reduce the fear, loneliness, and suffering of healthcare personnel and their patients. Even bringing a laptop to the bedside to share a personalized recorded music concert can add to the hope, relationship, emotional sharing, and comfort of patient and provider.

Psychology undergraduate students being trained in spatial reasoning and language processing while listening to Mozart's music in the background increased their speed of processing on both tasks on the third day.[47] That finding suggests that predictable improvement can be influenced by hearing music while processing spatial and cognitive data. Young, nonmusician university students that listened to music while encoding had better word recognition than peers who were kept in silence. The students listening to music also had enhanced activation of the parts of the brain associated with enhanced encoding and retrieval. This could be added to corporate settings to facilitate new learning with new and existing employees.

Ferreri and colleagues[48] had young, nonmusician university students listen to "If you see my mother" by Sidney Bechet. Enhanced activation of the dorsolateral prefrontal cortex correlated with enhanced encoding and retrieval in those listening to music compared to peers who were kept in silence during these tasks.

By providing structure, music can be an effective mnemonic which is increasingly effective when the music is familiar to the learner.[49] College students sometimes found benefits in listening to instrumental music in the background while studying. However, music with lyrics interfered with learning arithmetic, reading comprehension, and tasks which required visual and verbal learning.[50]

Nonmusician healthy individuals aged 60–84 were provided with the background music of Mozart's Eine Kleine Nachtmusik for its high arousal and positive emotion inducement; they responded with significant increase in their processing speed in comparison to the other three groups that had silence; white noise; or Mahler's Adagietto Symphony 5 which was included for its lower levels of arousal and inducement of negative emotions.[51] In that study, both Mozart and Mahler as background music were associated with better memory as compared to the groups tested with silence or white noise in the background. These scientists opined that the emotions induced by music facilitated the memory advantage in these individuals with, on average, 12.29 years of education.

Community-dwelling humans 65 years old and older were tested initially and again in six months. Those who participated weekly in an hour-long music-based exercise class during the six months had decreased anxiety and gained in the executive skill of resisting interference as compared to the control group.[52]

Japanese individuals 65 and older enjoyed physical exercise with musical accompaniment for one hour once a week completing 40 hours in one year.[53] With each session the exercise intensity gradually increased. Half of the group heard the music played in harmony with the exercise. The other half only heard percussion that kept the beat while the people read the lyrics without music while exercising. While both groups may have appreciated gains in psychomotor speed, only the music group had significant improvement in visuospatial function. These scientists believe that cognitive functioning in elders can be enhanced when music is combined with physical exercise.

Similarly, high tempo musical accompaniment played with physical exercise can reduce perceived exertion and increase duration of high intensity exercise sessions.[54] It can also influence faster heart rate recovery in healthy individuals recruited from a university community.

Another healthy response to hearing music is relief from painful emotions. This includes anxiety,[55] depression, and limited personal awareness of emotional status[56] even with patients hospitalized and on antidepressants for major depressive disorder.[57] Listening to music was associated with neurochemical changes associated with reduced stress; these could include increased release of dopamine and endogenous opioids as well as reduced cortisol and beta-endorphin[58]; receptive music compared to silence also was associated with maintaining healthier heart rate and blood pressure during stress.

Cortisol is a hormone which helps the body handle stress in part by regulating metabolism, reducing inflammation, improving the immune response, and improving memory. Music interventions have been associated with reducing psychological stress and stress markers such as cortisol.[59] Hearing music has an influence on human genes which "were shown to affect dopamine metabolism and to prevent neurodegeneration."[60] Community-dwelling elders, 65 or more years old had improvements in resisting interference, an important executive skill.[52]

Simply listening to music was associated with "increased life engagement and better health" in a dose-related way.[61] Among the over 20,000 Americans aged 50 or more who participated in the 2012 Health and Retirement Study and 2013 Consumption and Activities Mail Survey, 5.3% were high listeners who heard more than 28.6 hours of music per week. They reported more hours spent engaging in "sleeping, walking, participating in sports/ other activities, engaging in cognitively stimulating activities, and spending greater amounts of time engaged in prayers/meditations as compared to non-listeners." Average listeners also reported greater engagement in these activities than non-listeners. These researchers suggested including music in activities with elders to enhance aging with "direct health benefits."

At a six-month follow up with acute stroke patients afforded two months of up to an hour of listening to music,[62] individuals had improvement in "recovery of verbal memory and focused attention and reducing depression and confusion as well as increasing positive mood, relaxation, and motor activity." Their gains exceeded that of early post-stroke patients who had usual and customary care while listening to audiobooks. Better focused attention, verbal memory, and language skills correlated with grey matter volume increases and structural reorganization in frontal areas of the brains recovering from middle cerebral artery stroke.[62]

Six months after their stroke, patients had better memory for stories that were sung to them than the stories that were read to them.[63] Individuals having cardiac surgery who heard 30 minutes of relaxing music while resting prior to surgery were found to have increased oxytocin levels in postoperative measures compared to other post-cardiac surgery patients afforded ordinary hospital care without such music. Oxytocin is the hormone associated with positive emotions, social affiliation, and bonding.

Language and music share some cognitive resources, such as working memory.[64] However, musical learning and memory "engages brain networks that are distinct (or more widespread) than those involved in other types of episodic and semantic memory."[62] The brain "that engages in music is changed by engaging in music"[65,66]; rhythmic sounds of music provide "a physiological template for cueing the timing of movements" which could be accomplished with repetitions of the pairing to compensate and facilitate reshaping neuroplasticity to improve motor function which can be accomplished through rhythmic training of the auditory system for very brief periods resulting in efficient neural impact on motor systems.[67]

Care providers were trained in digital technology to create personalized music in the nonprofit organization, Music & Memory.[68] With listening to their familiar music, vast improvements in people suffering advanced dementia included: "(1) Enhanced swallowing mechanism with Music & Memory prior to dining; (2) decreased incidents of choking during mealtime; (3) improved nutritional status; (4) reduced weight loss; (5) reduced need for speech interventions; (6) enhanced quality of life."

While elders with dementia listened to familiar music, they were observed to have more eye movement; more eye contact; greater joy; more engagement; increased speech; and decreased sleep, movement, and dancing. Cognitive abilities music can enhance includes learning, memory, and speech. By activating the limbic system and other brain systems related to emotionality, music increases a sense of well-being. Music therapy enhances brain plasticity to the extent that it "can cure dementia" according to one research group.

Receptive music which improves mood and reduces stress could also reduce aging negative effects while improving the immune response through neurochemical changes.[58] These could include a decrease in interleukin-6 (IL-6), which can have both pro- and anti-inflammatory influence, as well as an increase in growth hormones; moreover, finding increased salivary immunoglobulin A (s-IgA) associated with receptive music is significant since s-IgA is a reliable measure of immune response in the mucosal system.

Elders might also have improved sleep onset, duration, and efficiency after hearing "smooth and wordless" music according to a review of studies conducted in five countries.[73] For some elders improved quality of sleep was realized only after about three weeks of listening to music.[74] Music with a soft volume, slow tempo between 60 and 80 beats per minute, and having a smooth melody has been considered "sedative music" which was more effective than rhythm-centered music to improve sleep quality with adults aged 60 and over who listened to music four or more weeks.[75]

Impact factors of listening to music may include the aesthetics of music; the tendency of receptive music to elicit movements from subtle to dancing; increased access to memories induced by listening to music, social affiliation; and relating to another person such as a therapist while listening to music. These can be strong positive reinforcement for hearing music.

In summary, since hearing music activates many cerebellar, subcortical, temporal, and frontal regions as well as multiple networks irrespective of age, musical experience, and training,[76] listening to music can be an evidence-based intervention throughout the life course driving neuroplasticity in a positive direction to enhance neurocognition for healthy brain aging as well as for modulating pathophysiology. Listening to music has shown significant positive purchase from in utero and premature birth through healthy adults and including elders with dementia at end of life. These have been observed in neurochemical, emotional, behavioral, and cognitive measures and have complex multifocal and multiple brain network involvement.

Neurochemical changes associated with receptive music[58] have included increases in oxytocin, dopamine, growth hormone, salivary immunoglobulin A, and endogenous opioids. Researchers also reported decreased interleukin-6 as well as decreased cortisol, the stress and arousal hormone.

Emotional benefits of receptive musical experiences include stress reduction[37] with a more stable heart rate[70,71]; less anxiety[52,55]; less depression and better awareness of emotions[56] even during inpatient treatment for major depression[57]; improved mood[62]; and more joy.[68,69] Shrock[46] is an advocate for music daily to reduce the fear, loneliness, and suffering of healthcare personnel and their patients. Even bringing a laptop to the bedside to share a personalized recorded music concert can add to the hope, relationship, emotional sharing, and comfort of patient and provider.

Behavioral gains associated with listening to music include better feeding in neonates[70,71]; better nutrition through safer feeding of elders with advanced dementia[68]; less agitation, more eye contact, and more talkativeness in adults with dementia[69]; improved motor function in patients who have a movement disorder[72]; increased engagement with exercise[53] and activities of life as well as better health[61] and improved sleep.[73-75]

Cognitive gains listening to music include improved verbal processing, learning, spatial reasoning, and increased speed of processing[47,48,53]; better memory[32,62,63,77] and gains in psychomotor speed and enhanced visuospatial functioning.[53] It is significant that all these executive skills are essential in preserving and enhancing cognitive function across the entire lifespan for healthy aging.

Educational level and small sample sizes are factors. However, these studies invite further research on potential applications across the entire lifespan in clinical and business populations with individuals showing prefrontal impairments, elders, or patients with Alzheimer's as well as with healthy individuals with the goal to increase human potential. Using artificial intelligence to develop wearables could increase this database in ways that simply hearing music could reduce the cost of healthcare, increase productivity, and promote prosocial behaviors for peaceful coexistence.

Active Music Making

Gathering around the piano to sing in four-part harmony occurred on our farm almost every week. That's why I believe I was listening to music in utero.

Beginning in infancy, music activities have enhanced cognitive and social skills.[78] Six-month-old infants and their parents participated in a six-month program of active music making, including percussion instruments. Infant improvements included earlier use of prelinguistic communication and a positive influence on parent–infant social interactions. Researchers opined that infant neuroplasticity is very responsive to musical experiences.

A review of recent research on music with children 3–12 years old[79] found benefits of music training which included "greater emotional intelligence, academic performance, and pro-social skills." These authors opined that music needs to be included in scholastic activities as an important subject as well as being an educational tool in other academic subjects. Six-year-old children who were given 36 weeks of free weekly training in keyboard or voice[80] showed more increase in Full Scale IQ than peers in drama or no lessons. Because it is very adaptable, this includes immediate feedback, and includes positive feedback of aesthetics and progress, indicating that training in music might be a preferred intervention for improving neurocognition (executive functions) "in a far-transfer" way[81] with "statistically significant far-transfer effects."[82]

In educational settings my family had music lessons with all of my siblings playing trombone, base horn, or another musical instrument in the high school band. Family gatherings to sing while my mother played piano were the practice sessions that led to the Shaffer Sisters Trio performing on stage, radio, and TV as amateurs. Fun for all!

The interface between singing and speech requires "binding lyrics and melody into a unified representation." This activates very complex and bilateral brain networks on fMRI and PET scans[62] and has major implications for clinical applications to enhance brain functions in congenital as well as acquired brain dysfunction. Musical training is an intervention of power which "may facilitate regional brain maturation" as seen in underprivileged children who were living in poverty during the two years of musical training that started when they were six to seven years old.[83] Activity in the cortex entrained to rhythm and was increased with more years in musical training, which "can be attributed to neural plasticity that accompanies many years of musical training."[84]

Children with an average age of 8.7 who played a musical instrument for more than 30 min a week scored higher on intellectual ability and verbal ability compared to children not playing a musical instrument.[85] Intensity of musical practice correlated with their axial diffusivity on diffusion tensor imaging. Researchers opined that "the relationship between musical practice and intellectual ability is related to the maturation of white matter pathways in the auditory motor system" suggesting that "musical training may be a means of improving cognitive and brain health during development."

Youth in the third and fourth grades in elementary school who had social skills deficits were given group music lessons. Prosocial skills and sympathy increased whether participants were voluntary or compelled to train suggesting "group music training facilitates the development of prosocial skills." Even with short-term group music training of 20 days,[86] four to six-year-old children measured higher on tests of intelligence. These intellectual enhancements were "correlated with changes in functional brain plasticity during an executive-function task" showing that "transfer of a high-level cognitive skill is possible in early childhood." Children 7–11 years old given a year of music training showed superior improvement in working memory.[87] Adolescents[88] had significant improvement in social skills after 10 weeks of a group music therapy program of singing, drumming, and instrumental improvisation. Teaching social skills through "song lyrics and improvisation emerged as salient interventions" for promoting prosocial skills.

Online questionnaire data gathered from 1779 choristers yielded reports of significant physical, spiritual, social, and emotional benefits from singing in a choir.[89] These benefits were significantly greater for professional singers. Music training being positively associated with increased volume in the parahippocampus and inferior frontal cortex suggests that this training affects "a circuit of brain regions involved in executive function, memory, language, and emotion."[90]

Listening to music, singing, and playing an instrument improved episodic memory in healthy adults.[77] Elders using musical improvisation improved their verbal memory.[91] After twelve 70-min bi-weekly sessions of improvisational vocal and instrumental exercises, elders in a residential setting with mild to moderate cognitive impairment appreciated significant improvement in cognitive skills in comparison to the control group that had 45-min sessions of gymnastic activities twice a week for 12 weeks and remained stable.[92] This music intervention "improved participants' general cognitive functions, selective attention, planning and logic skills, and abilities related to access and lexical retrieval." Making music together once a month in a music café was a "powerful medium to promote wellbeing for community-dwelling people living with dementia" and for their care partners.[93,94] During the COVID-19 pandemic, music improvisation online helped sustain and enhance community in "Our Virtual Tribe."[95]

Learning to play the piano requires integrating the movements of both hands into a temporal and aesthetic context. Repetition of practice requiring coordination of both hands integrates motor and sensory networks "as well as multimodal integration regions."[96] It may be that this repetition being reinforced, both immediately and aesthetically, is related to the transfer of gains to many cognitive domains such as perceptual speed, executive functioning, and working memory.

About 40 years after they had stopped instrumental music training of 4–14 years duration, adults aged 55 to 76 demonstrated the fastest neural timing.[97] Adults with less than four years of training in musical instruments had neural timing that was less rapid than the group with more training; however, their speed of neural timing exceeded those with no training. Professional musicians aged 65–90, as compared to nonmusician age peers, had better functional connectivity between brain hemispheres with "significantly greater accuracy in tactile interhemispheric transfer" of information and accuracy of response.[98] This is in line with research finding a larger corpus callosum in people with extensive musical training and adds to data indicating that neuroplasticity in aging is positively influenced by music training and is associated with the integration of brain networks.

The above is also in line with the Rush Memory and Aging Project. The 964 individuals without cognitive impairment had an average age of 78.7 years and an average of 14.6 years of education when they agreed to annual evaluation and brain autopsy at their death.[99] In the beginning of the study, they were asked if by the age of 18 they had foreign language instruction; if they had any music lessons; and if so, how many years of each. Over the course of about six years of annual examinations, 396 individuals developed mild cognitive impairment (MCI). The risk of MCI was about 30% less in those who before the age of 18 had had more than four years of foreign language training; the same finding was true of those who prior to 18 years of age had had more than four years of music lessons. When compared to people who had neither foreign language training nor music lessons before turning 18, the risk of MCI was about 60% less in individuals who had more than four years of both foreign language instruction and music lessons in their early years. It is unknown how much of this could have been mediated by the extent to which individuals used these skills during the intervening years between youth and being in the study.

My six siblings all played an instrument in the school band. Most played the trombone. The only instrument I was willing to practice was my voice. So, my music training was limited to being in a chorus or choir. In the Shaffer Sisters Trio, we sang on stage, radio, and TV as amateurs. Language training was French which I would hear myself think when I was in a foreign country and didn't know the language.

Healthy and cognitively intact adults 60–84 years old and naïve to music were given piano training.[100] Compared to their own baseline and to other elders who enjoyed different leisure activities, the elders given piano training enjoyed gains in attention, visuomotor tracking, processing speed, motor function, executive skills, positive emotions, subjective well-being, and some elements of quality of life. Adults between 64 and 76 years old had increased bilateral cortical thickness after six months of piano training, as compared to those listening to and learning about music, showing cortical plasticity in five auditory-related regions of the brains of older adults.[101] This training stabilized the microstructure of the white matter in the fornix.[102] Healthy adults without previous piano training who spent a year in piano lessons had improved fine motor control with the best improvement in their right hand.[103] These researchers opined that this shows that motor demand adaptations in practicing piano generalize to motor domains independent of the piano, a skill transfer. Older women who had 15 weeks of training in drumming and singing had better memory for verbal and visual information than the same-aged controls who had either language training or no training.[104]

After an 8-week piano playing protocol,[105] fMRI findings and neuropsychological tests of seven individuals with mild traumatic brain injury (TBI) provided "evidence for a causal relationship between musical training and reorganization of neural networks promoting enhanced cognitive performance." They noted significant differences in activation in the part of the brain that "regulates higher order cognitive processing, such as executive functions." Using three months of piano and drum training and musical improvisation, adults with TBI had improved executive functioning which was associated with grey matter structural neuroplasticity in prefrontal areas[106] and rebuilding white matter "structural connectome."[107]

Taking piano lessons is on my To Do List. Cannot be soon, however.

In a comprehensive review of studies that included data on 1757 individuals on several continents, elders with dementia who participated in music therapy "had positive effects on disruptive behavior and anxiety and a positive trend for cognitive function, depression and quality of life" as measured by the MMSE and Self-Administered Gerocognitive Exam.[108] Support for using interactive music intervention to enhance cognitive functioning in elders 65 and older with cognitive dysfunction was found in a review of studies that included 966 participants.[109] A review of recent studies with a total of 495 elders with "probable MCI or dementia" by clinical diagnosis, or whose MMSE scores were "between (and including) 13 and 26," also found improved cognitive functioning after "physically participating in" making music.[110]

Music is a complex and multisensory form of enrichment that has a positive influence on neuroplasticity in several regions of the brain because it requires integration of audiovisual information as well as appreciation of abstract rules.[11,12] Magnetoencephalography measures with individuals with an average age of 26.45 found that the anterior prefrontal cortex played a central role and that the neuroplastic response was greater in musicians with long-term training than was noted in those with short-term training.[113] After four months of piano lessons, people aged 60–84 years enjoyed improved mood as well as significant improvements in the cognitive skills of attention, control, motor function, visual scanning, and executive functioning.[100] A recent review[114] found music training associated with enhanced cognition in a variety of musical and nonmusical skills "spanning from executive functions to creativity."

In the realm of memory, an important factor may be the chunking built into musical forms in the phrasing of rhythm, harmony, and melody. That might explain why someone could have better memory retention for information related to music for longer duration and greater proportion than one's concurrent loss of memory in other realms. Among the critical roles of information communicated by music is its aesthetic engagement as well as significant effects on development, learning, and recovery of function.

Age-associated losses in "audio visual speech-to-noise perception" can be reduced with long-term music making.[115] Compared to nonmusicians on fMRI responses, older musicians had greater specific neural activity which was like that of nonmusicians who were younger. This was related to the intensity of their training. Frontal-parietal, visual-motor, and speech-motor activation was greater in older musicians with greater sensorimotor neural alignment. These researchers opined that processing audiovisual speech in noise benefits noted in long-term musicianship is a result of preserving youthful sensorimotor regions.

Professional musicians trained to play musical instruments provide an exceptional opportunity for research in neuroplasticity.[96] Continuous feedback from auditory, emotional and visual systems integrated with motor responses of practice that is repeated over their years of refining skills facilitates connections between and integration of many motor, auditory, and visual brain regions and networks. Researchers[116] opine "that training of this neural network may produce cross-modal effects on other behavioral and cognitive operations that draw on this network" and that this supports "plasticity-based training in preserving brain functions in the elderly." Studies of musical improvisation, because it seems to include real-time constraints on creativity as well as increased cooperation of far-reaching brain networks, could increase data on maintaining and increasing the positive plastic purchase of musical actions[117,118] which are associated with a "special state of mind, both amongst the performers and their listeners."[119]

Perhaps studies cited above shed light on why playing music many times each week significantly decreased the risk for dementia in a 21-year prospective study.[120] Compared to nonmusicians, musicians have better visuospatial memory and conflict resolution suggesting more resistance to interfering memories in musically skilled elders. Orchestral musicians also retained better gray matter density suggesting that their musical activities helped maintain their neuroplastic gains to enhance and preserve cognitive control skills in elders. Elders who had more than ten years of musical instrument training and practice scored higher than nonmusicians in "tests of naming, visuospatial memory, visuomotor speed, visuospatial sequencing, and cognitive flexibility" suggesting that high engagement in instrumental musical activities throughout life could preserve and enhance cognition through advanced age.[52]

In summary, researchers have reported that increases in Full Scale IQ[80] and enhanced intelligence[85,86] have been found after active music training. Active music training was associated with improved communication and social skills even as early as the first year of life[78] with some potential for the duration of a broad range of cognitive skills for about 40 years.[97] Cognitive skill improvements found with musical training include speed of processing, memory, attention, academic performance, verbal fluency, and factors of executive functioning. These neurocognitive gains have been associated with significant neuroplastic structural changes such as increased bilateral cortical thickness[101]; stabilized white brain matter microstructure[102]; grey brain matter structural changes in the prefrontal cortex[106]; and rebuilding white brain matter "structural connectome."[107]

Thus, active music training can result in multiple direct benefits to physical, emotional, social, neuroplastic, and neurocognitive healthy aging at any age. It is noteworthy that individuals who had four years of musical training prior to the age of 18 had a 30% reduced risk of dementia.[99] Reframed in Positive Psychology,[121] youth who had four years of music training prior to the age of 18 had a 30% higher probability of maintaining neuroplastic and neurocognitive strengths for a longer healthspan compared to youth without music training. Also noteworthy are the neuroplastic and neurocognitive purchases of making music in elders, healthy or impaired. Learning and motor "practice makes plasticity" in athletes and musicians.[122] These neuroplastic structural and functional gains can be realized simply through the joy of making music which can be culturally adapted as well as fun such that these neurocognitive benefits could add significant value in commerce, productivity, clinical care, career development, community welfare, and personal ideal health.

Dance

Hearing and making music, people tend to move, some dance. Around the piano on the farm of my youth we would move while singing. The weekly square dancing was great!

Dancing is a multifaceted influence on driving neuroplasticity in a positive direction. It is both exercise and an artform. Social factors are added when not dancing alone; these may include the impacts of human touch which were emphasized in Rescuing Hug and in the research of Marian Diamond described herein. Compared to nondancers, expert ballroom dancers had more empathy which was associated with years dancing with partners; this was also reflected in functional and structural MRI findings.[123] There was a positive correlation between empathy and gray brain matter volumes with long-term training in ballroom dance.

If their increased dancing exercise is aerobic, this could increase neuroplastic and neurocognitive benefits.[124–126] Although long-term frequent receptive and active music making are efficient ways to reduce the risk of dementia, frequent dancing adds complexity which could afford better long-term retention of neurocognitive functions and brain structure. Recent reviews and RCT studies support this perspective. Positive neuroplasticity has been noted in both grey and white brain matter after long-term dance training.[127] Dancing and drumming from 30 min to 1 hour weekly also resulted in health benefits.[128]

Music and dance involve brain networks that overlap[129] including brain regions involving emotion, action, and perception. Harmony, melody, and rhythm in dance and music actively sustain pleasure in a cycle inducing emotions, actions, and learning; activity is noted in "specific hedonic brain networks" which warrants research "concerning links between psychological processes and behaviour, human flourishing, and the concept of *eudaimonia*."

In his review of 12 studies of healthy people aged 59 or older, Nascimento[130] found that "dance practice was associated with an improvement in functional connectivity, cognitive performance, and increased brain volumes." With regular dancing of sufficient intensity, several regions of the brain benefited from the increased use of visual, auditory, motor, and executive functions required to move in space to rhythm and others in the environment. Improvements were noted in "cognition, attention, executive functions and memory"; neuroplastic changes can "significantly increase and also preserve the performance of the functional capacity (postural control, walking speed) of older adults, contributing to their autonomy and quality of life."

A small random controlled trial[131] with healthy people more than 60 years old compared a control group living "life as usual" to a 60 minute, 3 times a week, dance intervention group. Irish country, African, Greek, and other types of dances were included with medium physical intensity; enhanced brain activity on resting state fMRI with the dance intervention correlated with improved attention and executive functions.

Depression in people with MCI and dementia was reduced with dancing.[132] Individuals with dementia might have neuropsychiatric symptoms which can be addressed with medications; however, medications can have troublesome side effects whereas dance could be "a safe intervention for people living with dementia."[133]

During the COVID-19 lockdown in Shanghai, HIIT dance became popular. The more frequently the Chinese people were active in HIIT dance at home, the more likely they were to avoid depression during the lockdown.[134]

Patients with COVID-19 were assessed medically, mentally, and cognitively before and after three months of aerobic dance training. For positive impacts on cognition and speed of processing, key factors were frequency and intensity of training "suggesting that patients can delay the decline of their cognitive function through early aerobic dance intervention."[132]

Providing one hour of online dance which was self-selected provided health benefits mentally and socially.[135] These researchers concluded that, although dance classes in person might be optimal, with dance classes online people "can experience a connection between the body, mind, and community." Traditional dances of India improved balance and motor skills more than neuromuscular training in "children with Down syndrome."[136] The benefits of dance can also be facilitated remotely[137] which may be necessary for rural settings.

A review of five random control trials with 358 participants who had MCI found better "global cognition, attention, immediate and delayed recall, and visuospatial ability" after one to three weeks of Latin ballroom and aerobic dances with light to moderate intensity.[138] Researchers opined that increased neurotrophic factors, such as BDNF, as well as increased growth factors, could have mediated the improved neuroplasticity. Dances were culturally adapted, and participants danced with a partner or in groups adding social interaction to the equation which adds a socioemotional factor[139] that warrants being measured. Another review of 11 RCTs with 1412 healthy participants aged 55 and older supports the theory that dance improves function of global cognition as measured by the MMSE and MOCA. Larger improvements were noted in dancers with MCI.

Assessing the impact of dance on elders diagnosed with MCI,[140] reviewers only considered RCTs which included control groups. Neurocognitive purchases of the dance interventions included significant improvements in global cognition; physical function; memory; language; and visuospatial function.

Memory improvements associated with dance interventions with or without MCI might be related to the increased hippocampal volume that has been reported in dancers[141]; it might also reflect aerobic exercise effects of improved neuroplasticity in white matter in the brains of healthy elders.[126] Since music and dance share similar neural correlates,[127] mastery in each of these activities has shown positive neuroplastic purchase. The lack of information on the intensity of exercise in aerobic dance interventions was also noted[142] in the review of 5 RCTs with 842 elders with MCI; global cognitive functioning and memory improvements were significant, with some gains in executive function.

In summary, despite limitations such as sample size, heterogeneity of research design and populations studied, as well as limited data on dance intensity and/or duration, reviewers and researchers opined that providing dance activities for people could be one of the most effective interventions to enhance neuroplasticity for structural, neurochemical, functional, prosocial, and neurocognitive gains while potentially improving physical health. Dancing that is aerobic could increase neurogenesis. People are more likely to do what is fun, easy, portable, adapted to their culture, and available at low to no cost. Music and dance meet these parameters and can enhance neuroplasticity in ways that improve physical, emotional, neurocognitive, immunochemical, and social health to bring change within seconds that can endure decades potentially "stimulating social sustainability."[143] Music and dance can be provided remotely. Providing group dance events could be an important, inexpensive, and nonpharmacological way to bring these benefits to global citizens including remote rural settings.

On the farm of my youth, square dancing Saturday evenings was a regular event. Although the farm work afforded more than enough physical exercise, the social and emotional benefits of square dancing with friends were a welcome source of neurobics for the fun of it at end of every week. We all shared the language of love and laughter.

Polyglots Could Earn Lots

Another neurobic with potential for significant neuroplastic purchases is speaking more than one language. After years of studying French in public school, there was no real place to use it on the farm. Living as a Peace Corp dependent in Borneo and traveling around the world, I would hear myself thinking in French if I didn't know the language. My hesitancy to speak it in French speaking countries prompted another two years of studying French in college. Still don't use it so I do not qualify as a polyglot.

Children living in bilingual homes need to develop and cope with language control, language conflict, and switching between languages. Thus, bilingual research has studied executive functioning which may develop from these challenges.[144] In their review of 58 articles, which included 125 tasks, they excluded studies on children older than 12 years of age. Children six years of age and younger were termed in the critical stage for learning whereas between 6 and 12 years of age was considered post critical. In critical age children studies found significant bilingual advantages in executive functioning compared to children who spoke only one language. This bilingual advantage tended to diminish with the older children.

Individuals with bilingual experiences of a wide range had greater "volumes of the bilateral caudate nucleus and nucleus accumbens."[145] The volumes plateaued in bilinguals with the most experience.

Monolingual Swedish speaking five to six-year-old children performed better on visuospatial skills and expressive vocabulary than the same-age-group of bilinguals speaking Swedish and Finish.[146] On Full Scale IQ no differences were found. This small study with only 45 mono and 34 bilingual children emphasized the importance of assessing individuals in both home languages with young children.

Another small study assessed 35 Spanish-English bilinguals and 39 monolingual English-speaking preschoolers from poor homes.[147] Preschoolers who were bilingual were significantly more efficient in learning novel words; this was "mediated by short-term memory." These researchers opined that interventions that target short-term memory with these high-risk children might promote learning words.

Researchers used advanced technology to assess scans that had been taken on former patients in a memory clinic at a hospital where they had been diagnosed with mild cognitive impairment.[148] Of these scans of patients with MCI, they found more grey brain matter (GM) loss in the 23 bilingual patients than in the scans of the 24 monolingual MCI patients. This brain atrophy in bilinguals was noted in several areas of the brain involved in language and memory functions. The strongest predictors of GM atrophy were bilingualism and age. In addition to English, the most common other languages spoken were "Yiddish (26%), Hungarian (13%), Hebrew (8%), and Italian (8%)." They opined that the "neuroplastic changes to these areas induced by bilingualism may make them more robust even in the presence of brain atrophy." Despite, the greater cognitive reserve of bilinguals, "at some point, the accumulated neuropathology will become impossible for the system to accommodate, and the cognitive decline will be more rapid."

Also significant are the findings of the Rush Memory and Aging Project[99] which studied 964 people with an average age of 78.6 years and an average of 14.6 years of education. This study was referenced above for the neuroplastic benefits of four or more years of music lessons; they had a 30% higher probability of retaining their enhanced neuroplasticity for longer. Neurocognitive functioning was tested annually with 21 cognitive tests including measures of semantic memory, working memory, perceptual speed, episodic memory, and visuospatial ability. In this group only 124 had more than four years of foreign language instructions AND in music; *each* topic of instruction was associated with a "30% lower risk of MCI compared with no instruction." It is noteworthy that "these effects were additive, with risk of MCI approximately 60% lower with high levels of both" music and foreign language instruction as compared to no music or foreign language instruction before age 18. Music and second language use during the approximately 60 years between age 18 and enrolling in the Rush Memory and Aging Project is unknown. Another factor is that years of instruction were given by self-report. However, the reduced risk of 30% for each AND the additive finding of 60% lower risk of MCI in people with four or more years of both music and foreign language instruction prior to age 18 argues in favor of viewing bilingualism as another important neurobic.

These very significant research findings should influence the inclusion of four years of both music education and instruction in a foreign language in academic settings prior to age 18 and perhaps beyond. Using both across the lifespan is also likely to be ideal neurobics given research above.

Creativity

A theme repeated throughout this writing is that using skills enhances plasticity. Practice makes plasticity. Another constant theme is the need to improve memory across the lifespan. That raises questions about which types, categories, intensity, and duration of activities could be on our list of ways to increase memory. As described above, cognitive training on the UFOV program increased memory for the task that was trained.

Another neurobic worth considering could be creativity. Being creative requires several neurocognitive skills including memory.

Using machine learning to study brain neuroimaging, it was found that semantic memory functional patterns of connectivity predicted creativity in real life.[149] "Specifically, more efficient and denser functional connectivity between the default, control, salience, motor, and visual networks predicted a more integrated semantic memory structure." Further, this "predicted more creative behaviors."

A small study by Ahmed and colleagues[150] comparing age groups (30 aged 18–30 years old versus 30 aged 60–80 years old) found "evidence to show that healthy ageing confers an advantage in creative performance in terms of greater fluency, originality, and flexibility during idea generation." That fits with the description of Ramscar and colleagues[151] of "The Myth of Cognitive Decline." They attribute changes in performance on many psychometric tasks found with aging as reflecting "consequences of learning on information processing, and not cognitive decline." Having accumulated more acquired knowledge puts more and different demands on their memory search "which escalates as experience grows."

Among the very impressive examples of healthy aging with superior creativity would be John Kander,[152] the "longest working Broadway composer in history" according to the CBS News interview on April 23, 2023, on his 96th birthday, and the week that "a new Kander and Ebb musical" opened on Broadway. It's New York, New York, which was "very loosely based on the 1977 movie." With his partner, they wrote several of the songs in that movie. He describes sitting down at the piano and his fingers start playing. One of the new songs for the newest New York, New York that recently opened on Broadway, he wrote in 45 minutes.

Discussion

Beyond question in the realm of evidence-based interventions to drive neuroplasticity in a positive direction, this chapter on neurobics is one of my favorites for several reasons. My lifelong fascination with all things mental, social, and psychological was multiplied and magnified by observations of one individual I worked with as described in the early pages of this chapter.

After that individual had 15 years of math and reading remediation in special education settings which failed to bring those cognitive skills up to a functional level, a computerized math program became available. What wasn't described above was the positive reinforcement which is essential. Initially, money for time on task was given as positive reinforcement because people don't like to do what they are not good at. Eventually this individual agreed that money was no longer necessary because increases in accuracy and speed were sufficient as positive reinforcement for progress related to time on task. When the bonus scores for increased speed and accuracy that were provided by the computerized program rose into the thousands, that was *serious* positive reinforcement. But nothing could have been better than the many times thereafter across several decades when this person's speed and accuracy of *mental* math *FAR* exceeded that of MDs, PhDs, and others with any level of education.

Computer programs can adjust to a person's current level of functioning and provide positive reinforcement for speed and accuracy with greater accuracy than any human can. Adjusting to the current level of functioning is essential so the person using the program has sufficient challenge to stay engaged but not so much difficulty that they become discouraged and leave before making progress. Designed in this way, a computer software program can assist an individual in building skills that persist, perhaps generalize.

Although a case study, this individual is a case in point. Time on task with this computerized math program accomplished what 15 years of math remediation in the public school system could not.

Observing this case study put wind beneath my wings to intensify and broaden my study of all evidence-based computer-based cognitive training interventions that are associated with driving neuroplasticity in a positive direction. The research of Karlene Ball and colleagues on a computerized program to increase one's UFOV has been similarly inspiring. It is a joy to share the research on UFOV, very promising to read about the many initial benefits associated with use of that program, and impressive to see the multiple benefits found in participants even ten years later. It is likely that software programs that reinforce speed and accuracy are also building visual-motor coordination and some type of memory. Reading in a research article by Ball[6] that "these interventions have the potential to reverse age-related decline" is instructive, encouraging, and believable. You do not want to know how many times I used (or was that played with) the UFOV program.

Another reason this chapter is one of my favorites relates back to the perspective that education is rarely sufficient. People are most likely to do what is fun, easy, portable, free, and culturally adaptable. Music and dance neuroplasticity enhance!

The capacity to enhance neuroplasticity begins in utero and has been documented across the lifespan in healthy individuals as well as at the end of life with severe dementia. Guidelines in the NIH Music-Based Intervention Toolkit[153] created by a panel of interdisciplinary scientific experts considered "research on music and health across the lifespan." It "defines the building blocks" of music-based interventions for research and could be a valuable source of information for healthcare professionals and business settings.

Brain activity of infants indicates memory for music heard during months prior to birth. Six-month-old infants engaged in music making had increased prosocial and prelinguistic activities. Youth given music lessons increased scores on IQ testing. Elders with neurocognitive deficits given piano lessons enjoyed some return of mental functions. From hearing music to making music to movement that occurs naturally in response to music we can bring these neuroplastic benefits to all people of all ages.

This can be accomplished online and tailored to meet the factors important to the participants; we can also extend the reach. For example, Dr. Rudolph Tanzi[154] is a brilliant neuroscientist. It was in his lab that the first Alzheimer's gene was isolated. He is also a gifted musician as is evident at nlm.com/rudytanzi which provides an inexhaustible selection of the piano brilliance of Rudy Tanzi playing music he composed. This as background music could increase the neuroplasticity and productivity of workers on the job; calm the residents of a facility; aid the academician in their studies and writing; and enhance the neuroplasticity of those who enjoy this piano excellence.

Another beneficial source is PlayingForChange.org which has many invaluable videos they've created toward their goal of increasing peaceful coexistence. One of my favorites is The Weight: "Take the weight off Fanny.

Put the weight right on me." This video has musicians in many countries bringing joy to the World. MusicMendsMinds.org is another site for videos and virtual activities in their intergenerational efforts to improve the lives of individuals with dementia with the perspective that music is medicine.

It is worth repeating the findings of the Rush Memory and Aging Project that four or more years of music lessons prior to age 18 resulted in a 30% higher probability of retaining enhanced neuroplasticity for more years. It is also empowering to note that this same study found a 30% higher probability of longer lasting neuroplastic gains with four or more years of second language instruction prior to age 18. It is wonderful and noteworthy that "these effects were additive, with risk of MCI approximately 60% lower with high levels of both" music and foreign language instruction as compared to no music or foreign language instruction before age 18.

That's simultaneously heartwarming and empowering. It also means that schools must provide both music and second language instruction for all people.

The message of hope and empowerment is well stated by Gage and colleagues:[155] "Despite the dramatic reductions in hippocampus-dependent function that accompany advancing age, there is also striking evidence that even the aged brain retains a high level of plasticity." Thus, "one promising avenue to reach the goal of successful aging might be to boost and recruit this plasticity, which is the interplay between neural structure, function, and experience, to prevent age-related cognitive decline and age-associated comorbidities."

Neurobics: Newness and Challenge for Neurons. Bring it on! For me that will be more reading, writing, teaching, and thinking about evidence-based interventions for driving neuroplasticity in a positive direction. What will your neurobics be?

Your neurons require neurobics for the health of it. Whether by using computerized cognitive training to increase speed of processing, by music and dance, learning a foreign language, being on the internet, or pursuing a creative endeavor, complex new learning can add pleasure to the evidence-based activities that can drive your neuroplasticity in a positive direction. The message of hope and empowerment is captured by Gage and colleagues writing that there is "striking evidence that even the aged brain retains a high level of plasticity."

As stated earlier in this writing, we can celebrate that our brains are plastic. Provide your brain cells with copious newness and challenge for FUN and growth.

References

1. Katz L & Rubin M. (1999) In *Keep Your Brain Alive*. New York: Workman.
2. Mahncke H, Bronstone A & Merzenich MM (2006). Brain Plasticity and Functional Losses in the Aged: Scientific Bases for a Novel Intervention. *Prog Brain Res*, 157: 81–109.
3. Houillon A, Lorenz RC, Boehmer W, Rapp MA, Heinz A, Gallinat J, et al. (2013). The Effect of Novelty on Reinforcement Learning. *Prog Brain Res*, 202: 415–439. https://doi.org/10.1016/B978-0-444-62604-2.00021-6.
4. Abuhamden S & Csikszentmihalyi M (2012). The Importance of Challenge for the Enjoyment of Intrinsically Motivated, Goal-Directed Activities. *Pers Soc Psychol Bull*, 38: 317–330.
5. Shaffer J. (2016). Neuroplasticity and Clinical Practice: Building Brain Power for Health. Front Psychol, 7, 1118. https://doi.org/10.3389/fpsyg.2016.01118
6. Jobe JB, Smith DM, Ball KK, Tennstedt SL, Marsiske M, Willis SL, et al. (2001). ACTIVE: A Cognitive Intervention Trial to Promote Independence in Older Adults. *Control Clin Trials*, 22(4): 453–479.
7. Ball KK, Berch DB, Helmers KF, Jobe JB, Leveck MD, Mariske M, et al. (2002). Effects of Cognitive Training Interventions with Older Adults: A Randomized Controlled Trial. *JAMA*, 288(18): 2271–2281.
8. Ball KK, Ross LA, Roth DL & Edwards JD. (2013). Speed of Processing Training in the ACTIVE Study: How Much Is Needed and Who Benefits? *J Aging and Health*, 25(8S): 65S–84S.
9. Rebok GW, Ball KK, Guey LT, Jones RN, Kim H-Y, King JW, et al. (2014). Ten-Year Effects of the Advanced Cognitive Training for Independent and Vital Elderly Cognitive Training Trial on Cognition and Everyday Functioning in Older Adults. *J Am Geriatr Soc*, 62: 16–24. https://doi.org/10.1111/jgs.12607.
10. Edwards JD, Fausto BA, Tetlow AM, Crorna RT & Valdés EG. (2018). Systematic Review and Meta-analyses of Useful Field of View Cognitive Training. *Neurosci Biobehav Rev*, 84, 72–91. https://doi.org/10.1016/j.neubiorev.2017.11.004
11. Sprague BN, Ross LA & Ball KK. (2023). Does Cognitive Training Reduce Falls across Ten Years?: Data from the ACTIVE Trial. *Int J Environ Res Public Health*, 20: 4941. https://doi.org/10.3390/ ijerph20064941
12. Berry AS, Zanto TP, Clapp WC, Hardy JL, Delahunt PB, Mahncke HW, et al. (2010). The Influence of Perceptual Training on Working Memory in Older Adults. *PLoS ONE*, 5(7): e11537.
13. Silbert LC, Dodge HH, Lahna D, Promjunyakul NO, Austin D, Mattek N, et al. (2016). Less Daily Computer Use Is Related to Smaller Hippocampal Volumes in Cognitively Intact Elderly. *J Alzheimers Dis*. https://doi.org/10.3233/JAD-160079
14. Anderson-Hanley C, Arciero PJ, Brickman AM, Nimon JP, Okuma N, Westen SC, et al. (2012). Exergaming and Older Adult Cognition: A Cluster Randomized Clinical Trial. *Am J Prev Med*, 42(2): 109–119.
15. Merzenich MM. (2013). *Soft-Wired: How the New Science of Brain Plasticity Can Change Your Life*. San Francisco, CA: Parnassus Publishing.
16. Zhang H, Huntley J, Bhome R, Holmes B, Cahill J, Gould Rl, et al. (2019). Effect of Computerised Cognitive Training on Cognitive Outcomes in Mild

Cognitive Impairment: A Systematic Review and Meta-analysis. *BMJ Open*, 9: e027062. doi: 10.1136/bmjopen-2018-027062.

17. Wolinsky FD, Vander Weg MW, Howren MB, Jones MP & Dotson MM. (2014). The Effect of Cognitive Speed of Processing Training on the Development of Additional IADL Difficulties and the Reduction of Depressive Symptoms: Results from the IHAMS Randomized Controlled Trial. *J Aging Health.* https://doi.org/10.1177/0898264314550715

18. Edwards JD, Myers C, Ross LA, Roenker DL, Cissell GM, McLaughlin AM, et al. (2009). The Longitudinal Impact of Cognitive Speed of Processing Training on Driving Mobility. *The Gerontologist*, 49(4): 485–494.

19. Smith GE, Housen P, Yaffe K, Ruff R, Kennison RF, Mahncke HW, et al. (2009). A Cognitive Training Program Based on Principles of Brain Plasticity: Results from the Improvement in Memory with Plasticity-Based Adaptive Cognitive Training (IMPACT) Study. *J Am Geriatr Soc*, 57(4): 594–603. https://doi.org/10.1111/j.1532-5415.2008.02167.x

20. Zelinski EM, Spina LM, Yaffe K, Ruff R, Kennison RF, Mahncke HW, et al. (2011). Improvement in Memory with Plasticity-Based Adaptive Cognitive Training: Results of the 3-month Follow-Up. *J Am Geriatr Soc*, 59(2): 258–265.

21. Wolinsky FD, Vander Weg MW, Martin R, Unverzagt FW, Ball KK, et al. (2009). The ACTIVE Cognitive Training Interventions and the Onset of and Recovery from Suspected Clinical Depression. *J Gerontol*, 64A(4): 468–472.

22. Merzenich MM, Van Vleet TM & Nahum M. (2014). Brain Plasticity-based Therapeutics. Front Hum Neurosci, June 27; 8: 385. https://doi.org/10.3389/fnhum.2014.00385. eCollection 2014.

23. Gajewski PD & Falkenstein M. (2016) Physical Activity and Neurocognitive Functioning in Aging - A Condensed Updated Review. *Eur Rev Aging Phys Act*, 13: 1. doi: 10.1186/s11556-016-0161-3.

24. Jessberger S & Gage FH. (2014). Adult Neurogenesis: Bridging the Gap between Mice and Humans. Cell Press, article in press. https://doi.org/10.1016/j.tcb.2014.07.003

25. Ge S, Yang C, Hsu K, Ming G & Song H. (2007). A Critical Period for Enhanced Synaptic Plasticity in Newly Generated Neurons of the Adult Brain. *Neuron*, 54(4): 559–566.

26. Dranovsky A, Picchini AM, Moadel T, Sisti AC, Yamada A, Kimura S, et al. (2011). Experience Dictates Stem Cell Fate in the Adult Hippocampus. *Neuron*, 70(5): 908–923.

27. Kempermann G, Gast D & Gage FH. (2002). Neuroplasticity in Old Age: Sustained Fivefold Induction of Hippocampal Neurogenesis by Long-Term Environmental Enrichment. *Ann Neurol*, 52: 135–143.

28. Malkasian D & Diamond MC. (1971). The Effect of Environmental Manipulation on the Morphology of the Neonatal Rat Brain. *Int J Neurosci*, 2: 161–170.

29. Uylings HBM, Kuypers K, Diamond MC & Veltman WAM. (1978). The Effects of Differential Environments on Plasticity of Cortical Pyramidal Neurons in Adult Rats. *Exp Neurol*, 68: 158–170.

30. Diamond M, Johnson R, Protti A, Ott C & Kajisa L. (1984). Plasticity in the 904-Day-Old Male Rat Cerebral Cortex. *Exp Neurol*, 87: 309–317.

31. Deshpande A, Bergami M, Ghanem A, Conzelmann K-K, Lepier A, Götz M et al. (2013). Retrograde Monosynaptic Tracing Reveals the Temporal Evolution of Inputs onto New Neurons in the Adult Dentate Gyrus and Olfactory Bulb.*Proc Natl Acad Sci U S A*, 110(12): E1152–E1161.
32. Partanen E, Kujala T, Tervaniemi, M & Huotilainen M. (2013) Prenatal Music Exposure Induces Long-Term Neural Effects. PLoS ONE, 8: e78946. https://doi.org/10.1371/journal.pone.0078946
33. Movalled K, Sani A, Nikniaz L & Ghojazadeh M. (2023). The Impact of Sound Stimulations During Pregnancy on Fetal Learning: A Systematic Review. BMC Pediatr, 23: 183. https://doi.org/10.1186/s12887-023-03990-7
34. Särkämö T. (2018). Music for the Ageing Brain: Cognitive, Emotional, Social, and Neural Benefits of Musical Leisure Activities in Stroke and Dementia. *Dementia*, 17: 670–685.
35. Wallmark Z, Deblieck C & Iacoboni M. (2018) Neurophysiological Effects of Trait Empathy in Music Listening. *Front Behav Neurosci*, 12: 66. https://doi.org/10.3389/fnbeh.2018.00066
36. Detmer MR & Whelan ML. (2017) Music in the NICU: The Role of Nurses in Neuroprotection. Neonatal Netw, 36: 213–217. DOI: 10.1891/0730-0832.36.4.213
37. Anderson D & Patel AD. (2018) Infants Born Preterm, Stress, and Neurodevelopment in the Neonatal Intensive Care Unit: Might Music Have an Impact? *Dev Med Child Neurol*, 60: 256–266. https://doi.org/10.1111/dmcn.13663
38. Bos M, van Dokkum NH, Ravensbergen A, Kraft KE, Bos AF & Jaschke AC. (2021). Pilot Study Finds That Performing Live Music Therapy in Intensive Care Units May Be Beneficial for Infants' Neurodevelopment. *Acta Paediatr*, 110: 2350–2351. https://doi.org/10.1111/apa.15867
39. Palazzi A, Meschini R & Piccinini, CA. (2021). NICU Music Therapy Effects on Maternal Mental Health and Preterm Infant's Emotional Arousal. *Infant Ment Health J*, 42: 672–689. https://doi.org/10.1002/imhj.21938
40. Mikulis N, Inder TE & Erdei C. (2021). Utilising Recorded Music to Reduce Stress and Enhance Infant Neurodevelopment in Neonatal Intensive Care Units. Acta Paediatr, 110: 2921–2936. https://doi.org/10.1111/apa.15977.
41. Loukas S, Lordier L, Meskaldji D, Filippa M, de Almeida JS, Van De Ville D & Hüppi PS. (2022). Musical Memories in Newborns: A Resting-State Functional Connectivity Study. *Hum. Brain Mapp*, 43: 647–664. https://doi.org/10.1002/hbm.25677
42. Yue W, Han X, Luo J, Zeng Z & Yang M. (2021). Effect of Music Therapy on Preterm Infants in Neonatal Intensive Care Unit: Systematic Review and Meta-analysis of Randomized Controlled Trials. *J Adv Nurs*, 77: 635–652. https://doi.org/10.1111/jan.14630
43. Almeida JS, Baud O, Fau S, Barcos-Munoz F, Courvoisier S, Lordier L, et al. (2023). Music Impacts Brain Cortical Microstructural Maturation in Very Preterm Infants: A Longitudinal Diffusion MR Imaging Study. *Dev Cogn Neurosc*, 61: 101254. https://doi.org/10.1016/j.dcn.2023.101254
44. Palaskar P, Ramekar SD, Sant N, et al. (2023) Ideal Mode of Auditory Stimulation in Preterm Neonates in Neonatal Intensive Care Unit: A Systematic Review. *Cureus*, 15(2): e34496. https://doi.org/10.7759/cureus.34496

45. Kuuse A-K, Paulander A-S & Eulau L. (2023) Characteristics and Impacts of Live Music Interventions on Health and Wellbeing for Children, Families, and Health Care Professionals in Paediatric Hospitals: A Scoping Review. *Int J Qual Stud Heal*, 18: 2180859. https://doi.org/10.1080/17482631.2023.2180859

46. Shrock CL, (2023). A Daily Dose of Music. *Acad Pediatr* https://doi.org/10.1016/j.acap.2023.05.005

47. Angel LA, Polzella DJ & Elvers GC. (2010). Background Music and Cognitive Performance. *Percept Mot Ski*, 110: 1059–1064.

48. Ferreri L, Aucouturier J, Muthalib M, Bigand E & Bugaiska A. (2013). Music Improves Verbal Memory Encoding while Decreasing Prefrontal Cortex Activity: An fNIRS Study. *Front Hum Neurosci*, 7: Article 779. https://doi.org/10.3389/fnhum.2013.00779

49. Derks-Dijkman MW, Schaefer RS & Kessels RPC. (2023). Musical Mnemonics in Cognitively Unimpaired Individuals and Individuals with Alzheimer's Dementia: A Systematic Review. *Neuropsychol* Rev. https://doi.org/10.1007/s11065-023-09585-4

50. Souza AS & Barbosa LCL. (2023). Should We Turn off the Music? Music with Lyrics Interferes with Cognitive Tasks. *J Cognition*, 6(1): 24, pp. 1–18. https://doi.org/10.5334/joc.273

51. Bottiroli S, Rossi A, Russo R, Vecchi T & Cavallini E. (2014). The Cognitive Effects of Listening to Background Music on Older Adults: Processing Speed Improves with Upbeat Music, While Memory Seems to Benefit from Both Upbeat and Downbeat Music. *Front Aging Neurosci*, 6: Article 284. https://doi.org/10.3389/fnagi.2014.00284

52. Hars M, Herrmann FR, Gold G, Rizzoli R & Trombetti A. (2014). Effect of Music-Based Multitask Training on Cognition and Mood in Older Adults. *Age Ageing*, 43: 196–200. https://doi.org/10.1093/ageing/aft163

53. Satoh M, Ogawa J, Tokita T, Nakaguchi N, Nakao K, Kida H et al. (2014). The Effects of Physical Exercise with Music on Cognitive Function of Elderly People: Mihama-Kiho Project. *PLoS ONE*, 9(4): e95230. https://doi.org/10.1371/journal.pone.0095230

54. Maddigan ME, Sullivan KM, Halperin I, Basset FA & Behm DG. (2019). High Tempo Music Prolongs High Intensity Exercise. *PeerJ*, 6: e6164. https://doi.org/10.7717/peerj.6164

55. Gurbuz-Dogan RN, Ali B, Candy B & King M. (2021). The Effectiveness of Sufi Music for Mental Health Outcomes. A Systematic Review and Meta-Analysis of 21 Randomised Trials. *Complem Thera Med*, 57: 102664.

56. Erkkila J, Punkanen M, Fachner J, Ala-Ruona E, Pontio I, Tervaniemi M, Vanhala M & Gold C. (2011). Individual Music Therapy for Depression: Randomised Controlled Trial. *BJPsych*, 199: 132–139. https://doi.org/10.1192/bjp.bp.110.085431

57. Hsu W-C & Lai H-L. (2004). Effects of Music on Major Depression in Psychiatric Inpatients. *Arch Psychiatr Nurs*, 18: 193–199. https://doi.org/10.1016/j.apnu.2004.07.007

58. Chanda ML & Levitin DJ. (2013). The Neurochemistry of Music. *Trends Cogn Sci*, 17: 179–193. https://doi.org/10.1016/j.tics.2013.02.007

59. Wong MM, Tahir T, Wong MM, Baron A & Finnerty R. (2021). Biomarkers of Stress in Music Interventions: A Systematic Review. *J Music Ther*, 58: 241–277. https://doi.org/10.1093/jmt/thab003

60. Nair PS, Raijas P, Ahvenainen M, Philips AK, Ukkola-Vuoti L & Järvelä I. (2020), I. Music-Listening Regulates Human microRNA Expression. *Epigenetics*, 16: 554–566. https://doi.org/10.1080/15592294.2020.1809853

61. Kaufmann CN, Montross-Thomas LP & Griser S. (2017). Increased Engagement with Life: Differences in the Cognitive, Physical, Social, and Spiritual Activities of Older Adult Music Listeners. *Gerontol*, 58: 270–277. https://doi.org/10.1093/geront/gnw192

62. Särkämö T. & Sihvonen AJ (2018). Golden Oldies and Silver Brains: Deficits, Preservation, Learning, and Rehabilitation Effects of Music in Ageing-Related Neurological Disorders. *Cortex*, 109: 104e123. https://doi.org/10.1016/j.cortex.2018.08.034

63. Leo V, Sihvonen AJ, Linnavalli T, Tervaniemi M, Laine M, Soinila S et al. (2018) Sung Melody Enhances Verbal Learning and Recall After Stroke. *Ann NY Acad Sci*, 1423: 296–307. https://doi.org/10.1111/nyas.13624

64. Atherton RP, Chrobak QM, Rauscher FH, Karst AT, Hanson MD, Steinert SW, et al. (2018). Shared Processing of Language and Music. *Exp Psychol*, 65: 40–48. https://doi.org/10.1027/1618-3169/a000388

65. Thaut MH. (2005) The Future of Music in Therapy and Medicine. *Ann NY Acad Sci*, 1060: 303–308. https://doi.org/10.1196/annals.1360.023

66. Thaut MH. (2021). Advances in the Role of Music in Neurorehabilitation: Addressing Critical Gaps in Clinical Applications. *NeuroRehabilitation*, 48: 153. https://doi.org/10.3233/NRE-208010

67. Crasta JE, Thaut MH, Anderson CW, Davies PL & Gavin WJ. (2018). Auditory Priming Improves Neural Synchronization in Auditory-Motor Entrainment. *Neuropsychologia*, 117: 102–112. https://doi.org/10.1016/j.neuropsychologia.2018.05.017

68. Cohen D, Post SG, Lo A, Lombardo R & Pfeffer B. (2018). "Music & Memory" and Improved Swallowing in Advanced Dementia. *Dementia*, 19: 195–204. https://doi.org/10.1177/1471301218769778

69. Ihara ES, Tompkins CJ, Inoue, M & Sonneman S. (2018). Results from a Person-Centered Music Intervention for Individuals Living with Dementia. *Geriatr Gerontol Int*, 19: 30–34. https://doi.org/10.1111/ggi.13563

70. Arnon S, Diamant C, Bauer S, Regev R, Sirota G & Litmanovitz I. (2014) Maternal Singing During Kangaroo Care Led to Autonomic Stability in Preterm Infants and Reduced Maternal Anxiety. *Acta Paediatr*, 103: 1039–1044. https://doi.org/10.1111/apa.12744

71. Pineda R, Guth R, Herring A, Reynolds L, Oberle S & Smith J. (2017). Enhancing Sensory Experiences for Very Preterm Infants in the NICU: An Integrative Review. *J Perinatol*, 37: 323–332. https://doi.org/10.1038/jp.2016.179

72. Thaut MH, McIntosh GC & Ehoemberg V. (2015). Neurobiological Foundations of Neurologic Music Therapy: Rhythmic Entrainment and the Motor System. *Front* Psychol, 5: 1185–1185. https://doi.org/10.3389/fpsyg.2014.01185

73. Wang C, Li G, Zheng L, Meng X, Meng Q, Wang S, et al. (2021) Effects of Music Intervention on Sleep Quality of Older Adults: A Systematic Review and Meta-analysis. *Complement Ther Med*, 59: 102719. https://doi.org/10.1016/j.ctim.2021.102719

74. Tang YW, Teoh SL, Yeo JHH, Ngim CF, Lai NM, Durrant SJ & Lee SWH. (2021). Music-Based Intervention for Improving Sleep Quality of Adults without Sleep Disorder: A Systematic Review and Meta-analysis. *Behav Sleep Med*, 20: 241–259. https://doi.org/10.1080/15402002.2021.1915787

75. Chen C, Tung H, Fang C, Wang J, Ko N, Chang Y & Chen Y. (2021). Effect of Music Therapy on Improving Sleep Quality in Older Adults: A Systematic Review and Meta-analysis. *J Am Geriatr Soc*, 69: 1925–1932. https://doi.org/10.1111/jgs.17149

76. Chan MMY & Han YMY. (2022). The Functional Brain Networks Activated by Music Listening: A Neuroimaging Meta-Analysis and Implications for Treatment. *Neuropsychol*, 36(1): 4–22. https://doi.org/10.1037/neu0000777

77. Rouse HJ, Jin Y, Hueluer G, Huo MA, Bugos J, Veal B, et al. (2021). Association Between Music Engagement and Episodic Memory Among Middle-Aged and Older Adults: A National Cross-Sectional Analysis. *Journals Gerontol Ser B Psychol Sci Soc Sci*, 77: 558–566. https://doi.org/10.1093/geronb/gbab044

78. Gerry D, Unrau A & Trainor L. (2012). Active Music Classes in Infancy Enhance Musical, Communicative and Social Development. *Dev Sci*, 15: 398–407.

79. Blasco-Magraner J, Bernabe-Valero G, Marín-Liébana P & Moret-Tatay C. (2021). Effects of the Educational Use of Music on 3- to 12-Year-Old Children's Emotional Development: A Systematic Review. *Int J Environ Res Public Health*, 18: 3668. https://doi.org/10.3390/ijerph18073668

80. Schellenberg EG. (2004). Music Lessons Enhance IQ. *Psychol Sci*, 15: 511–514. https://doi.org/10.1111/j.0956-7976.2004.00711.x

81. Degé F (2021) Music Lessons and Cognitive Abilities in Children: How Far Transfer Could Be Possible. *Front Psychol*, 11: 557807. https://doi.org/10.3389/fpsyg.2020.557807

82. Bigand E & Tillmann B. (2022). Near and Far Transfer: Is Music Special? *Mem Cogn*, 50: 339–347. https://doi.org/10.3758/s13421-021-01226-6

83. Habibi A, Damasio A, Ilari B, Veiga R, Joshi A, Leahy RM et al. (2017). Childhood Music Training Induces Change in Micro and Macroscopic Brain Structure: Results from a Longitudinal Study. Cereb Cortex, 28: 4336–4347. https://doi.org/10.1093/cercor/bhx286

84. Harding EE, Sammler D, Henry MJ, Large EW & Kotz SA. Cortical Tracking of Rhythm in Music and Speech. NeuroImage, 185: 96–101. https://doi.org/10.1016/j.neuroimage.2018.10.037

85. Loui P, Raine LB, Chaddock-Heyman L, Kramer AF & Hillman CH. (2019). Musical Instrument Practice Predicts White Matter Microstructure and Cognitive Abilities in Childhood. *Front* Psychol, 10: 1198. https://doi.org/10.3389/fpsyg.2019.01198

86. Moreno S, Bialystok E, Barac R, Schellenberg EG, Cepeda NJ & Chau T. (2011). Short-Term Music Training Enhances Verbal Intelligence and Executive Function. *Psychol Sci*, 22: 1425–1433. https://doi.org/10.1177/0956797611416999

87. Nie P, Wang C, Rong G, Du B, Lu J, Li S et al. (2022). Effects of Music Training on the Auditory Working Memory of Chinese-Speaking School-Aged Children: A Longitudinal Intervention Study. Front Psychol, 12: 770425. https://doi.org/10.3389/fpsyg.2021.770425

88. Pasiali V & Clark C. (2018) Evaluation of a Music Therapy Social Skills Development Program for Youth with Limited Resources. *J Music Ther*, 55: 280–308. https://doi.org/10.1093/jmt/thy007

89. Moss H, Lynch J & Donoghue JO (2018). Exploring the Perceived Health Benefits of Singing in a Choir: An International Cross-sectional Mixed-Methods Study. *Perspect Public Health*, 138: 160–168. https://doi.org/10.1177/1757913917739652

90. Chaddock-Heyman L, Loui P, Weng T, Weisshappel R, McAuley E & Kramer AF. (2021). Musical Training and Brain Volume in Older Adults. Brain Sci, 11, 50. https://doi.org/10.3390/brainsci11010050

91. Abrahan VD, Shifres F & Justel N. (2020). Impact of Music-Based Intervention on Verbal Memory: An Experimental Behavioral Study with Older Adults. Cogn Process, 22, 117–130. https://doi.org/10.1007/s10339-020-00993-5

92. Biasutti M & Mangiacotti A. (2017). Assessing a Cognitive Music Training for Older Participants: A Randomised Controlled Trial. *Int J Geriatr Psychiatry*, 33: 271–278. https://doi.org/10.1002/gps.4721

93. Smith SK, Innes A & Bushell S. (2021). Music-Making in the Community with People Living with Dementia and Care-Partners – 'I'm Leaving Feeling on Top of the World'. *Health Soc Care Community*, 30: 114–123. https://doi.org/10.1111/hsc.13378

94. Mittelman MS & Papayannopoulou PM. (2018). The Unforgettables: A Chorus for People with Dementia with Their Family Members and Friends. *Int Psychogeriatrics*, 30: 779–789. https://doi.org/10.1017/s1041610217001867

95. MacDonald R, Burke R, De Nora T, Donohue MS & Birrell R. (2021). Our Virtual Tribe: Sustaining and Enhancing Community via Online Music Improvisation. Front Psychol, 11: 623640. https://doi.org/10.3389/fpsyg.2020.623640

96. Schlaug G. (2015). Musicians and Music Making as a Model for the Study of Brain Plasticity. Prog Brain Res, 217: 37–55. https://doi.org/10.1016/bs.pbr.2014.11.020

97. White-Schwoch T, Carr KW, Anderson S, Strait DL & Kraus N. (2013). Older Adults Benefit from Music Training Early in Life: Biological Evidence for Long-Term Training-Driven Plasticity. *J Neurosci*, 33: 17667–17674. https://doi.org/10.1523/jneurosci.2560-13.2013

98. Piccirilli M, Palermo MT, Germani A, Bertoli ML, Ancarani V, Buratta L, et al. (2020). Music Playing and Interhemispheric Communication: Older Professional Musicians Outperform Age-Matched Non-Musicians in Fingertip Cross-Localization Test. *J Int Neuropsychol Soc*, 27: 282–292. doi.org/10.1017/s1355617720000946

99. Wilson RS, Boyle PA, Yang J, James BD & Bennett DA. (2014). Early Life Instruction in Foreign Language and Music and Incidence of Mild Cognitive Impairment. Neuropsychology. https://doi.org/10.1037/neu0000129

100. Seinfeld S, Figueroa H, Ortiz-Gil J & Sanchez-Vives MV. (2013). Effects of Music Learning and Piano Practice on Cognitive Function, Mood and Quality of Life in Older Adults. *Front Psychol*, 4: Article 810. https://doi.org/10.3389/fpsyg.2013.00810

101. Worschech F, Altenmüller E, Jünemann K, Sinke C, Krüger THC, Scholz DS, et al. (2022). Evidence of Cortical Thickness Increases in Bilateral Auditory Brain Structures Following Piano Learning in Older Adults. *Ann NY Acad Sci*. https://doi.org/10.1111/nyas.14762

102. Junemann K, Marie D, Worschech F, Scholz DS, Grouillero F, Kliegel M, et al. (2022). Six Months of Piano Training in Healthy Elderly Stabilizes White Matter Microstructure in the Fornix, Compared to an Active Control Group. *Front Aging* Neurosci, 14: 817889. https://doi.org/10.3389/fnagi.2022.817889

103. Worschech F, James CE, Jünemann K, Sinke C, Krüger THC, Scholz DS, et al. (2023). Fine Motor Control Improves in Older Adults After 1 Year of Piano Lessons: Analysis of Individual Development and Its Coupling with Cognition and Brain Structure. *Eur J Neurosci*: 1–22. https://doi.org/10.1111/ejn.16031

104. Degé F & Kerkovius K. (2018) The Effects of Drumming on Working Memory in Older Adults. *Ann NY Acad Sci*, 1423: 242–250. https://doi.org/10.1111/nyas.13685

105. Vik BMD, Skeie GO, Vikane E & Specht K. (2018). Effects of Music Production on Cortical Plasticity within Cognitive Rehabilitation of Patients with Mild Traumatic Brain Injury. Brain Inj, 32: 634–643. https://doi.org/10.1080/02699052.2018.1431842

106. Siponkoski S.-T, Martínez-Molina N, Kuusela L, Laitinen S, Holma M, Ahlfors M, et al. (2020). Music Therapy Enhances Executive Functions and Prefrontal Structural Neuroplasticity after Traumatic Brain Injury: Evidence from a Randomized Controlled Trial. *J Neurotrauma*, 37: 618–634. https://doi.org/10.1089/neu.2019.6413

107. Sihvonen AJ, Siponkoski S-T, Martinez-Molina N, Laitinen S, Holma M, Ahlfors M, et al. (2021). The Use of Artificial Neural Networks to Predict the Physicochemical Characteristics of Water Quality in Three District Municipalities, Eastern Cape Province, South Africa. *Int J Environ Res* Public Health, 18: 5248.

108. Zhang Y, Cai J, An L, Hui F, Ren T, Ma H & Zhao Q. (2017). Does Music Therapy Enhance Behavioral and Cognitive Function in Elderly Dementia Patients? A Systematic Review and Meta-analysis. *Aging Res Rev*, 35: 1–11. https://doi.org/10.1016/j.arr.2016.12.003

109. Xu B, Sui Y, Zhu C, Yang X, Zhou J, Li L, Ren L & Wang X. (2017). Music Intervention on Cognitive Dysfunction in Healthy Older Adults: A Systematic Review and Meta-analysis. *Neurol Sci*, 38: 983–992. https://doi.org/10.1007/s10072-017-2878-9

110. Dorris JL, Neely S, Terhorst L, VonVille HM & Rodakowski J. (2021). Effects of Music Participation for Mild Cognitive Impairment and Dementia: A Systematic Review and Meta-analysis. *J Am Geriatr Soc*, 69: 2659–2667. https://doi.org/10.1111/jgs.17208

111. Paraskevopoulos E, Kuchenbuch A, Herholz SC & Pantev C. (2012). Musical Expertise Induces Audiovisual Integration of Abstract Congruency Rules. *J Neurosci*, 32(50): 18196–18203. doi: 10.1523/JNEUROSCI.1947-12.2012.

112. Kuchenbuch A, Paraskevopoulos E, Herholz SC & Pantev C. (2014). Audio-Tactile Integration and the Influence of Musical Training. *PLoS ONE*, 9(1): e85743. DOI: 10.1371/journal.pone.0085743

113. Paraskevopoulos E, Kuchenbuch A, Herholz SC & Pantev C. (2014). Multisensory Integration during Short-Term Music Reading Training Enhances Both Uni- and Multisensory Cortical Processing. *J Cogn Neuroscience*, 26(10): 2224–2238. https://doi.org/10.1162/jocn_a_00620

114. Benz S, Sellaro R, Hommel B & Colzato LS. (2016). Music Makes the World Go Round: The Impact of Musical Training on Non-musical Cognitive Functions—A Review. Front Psychol, 6, Article 2023. DOI: 10.3389/fpsyg.2015.02023

115. Zhang L, Wang X, Alain C & Du Y. (2023). Successful Aging of Musicians: Preservation of Sensorimotor Regions Aids Audiovisual Speech-in-Noise Perception. *Sci Adv*, 9: eadg7056.

116. Wan CY & Schlaug G. (2010). Music Making as a Tool for Promoting Brain Plasticity across the Life Span. Neuroscientist, 16: 566–577. https://doi.org/10.1177/1073858410377805

117. Loui P. (2018). Rapid and Flexible Creativity in Musical Improvisation: Review and a Model. *Ann NY Acad Sci*, 1423: 138–145. https://doi.org/10.1111/nyas.13628

118. Beaty RE. The Neuroscience of Musical Improvisation. *Neurosci Biobehav Rev*, 51: 108–117. https://doi.org/10.1016/j.neubiorev.2015.01.004

119. Dolan D, Jensen HJ, Mediano PAM, Molina-Solana M, Rajpal H, Rosas F & Sloboda JA. (2018). The Improvisational State of Mind: A Multidisciplinary Study of an Improvisatory Approach to Classical Music Repertoire Performance. Front Psychol, 9: 1341. https://doi.org/10.3389/fpsyg.2018.01341

120. Verghese J, Lipton RB, Katz MJ, Hall CB, Derby CA, Kuslansky G, et al. (2003). Leisure Activities and the Risk of Dementia in the Elderly. *New Engl J Med*, 348: 2508–2516. https://doi.org/10.1056/nejmoa022252

121. Peterson, C. (2008) What Is Positive Psychology, and What Is It Not? Psychol Today, May 16. https://www.psychologytoday.com/us/blog/the-good-life/200805/what-is-positive-psychology-and-what-is-it-not? (accessed online Aug 12, 2023).

122. Kweon J, Vigne MM, Jones RN, Carpenter LL & Brown JC. (2023) Practice Makes Plasticity: 10-Hz rTMS Enhances LTP-like Plasticity in Musicians and Athletes. *Front Neural Circuits*, 17: 1124221. https://doi.org/10.3389/fncir.2023.1124221

123. Wu X, Lu X, Zhang H, Wang X, Kong Y & Hu L. (2023). The Association Between Ballroom Dance Training and Empathic Concern: Behavioral and Brain Evidence. *Hum Brain Mapp*, 44: 315–326. https://doi.org/10.1002/hbm.26042.

124. Fari G, Lunetti P, Pignatelli G, Raele MV, Cera A, Mintrone G, et al. (2021). The Effect of Physical Exercise on Cognitive Impairment in Neurodegenerative Disease: From Pathophysiology to Clinical and Rehabilitative Aspects. *Int J Mol Sci*, 22: 11632. doi: 10.3390/ijms222111632.

125. Pereira AC, Huddleston DE, Brickman AM, Sosunov AA, Hen R, McKhann GM, et al. (2007). An in Vivo Correlate of Exercise-Induced Neurogenesis in the Adult Dentate Gyrus. *Proc Natl Acad Sci U S A*, 104(13): 5638–5643.

126. Colmenares AM, Voss MW, Fanning J, Salerno EA, Gothe NP, Thomas ML, et al. (2021). White Matter Plasticity in Healthy Older Adults: The Effects of Aerobic Exercise. NeuroImage, 239: 118305. https://doi.org/10.1016/j.neuroimage.2021.118305

127. Karpati FJ, Giacosa C, Foster NE, Penhune VB & Hyde KL. (2017). Dance and Music Share Gray Matter Structural Correlates. Brain Res, 1657: 62–73. https://doi.org/10.1016/j.brainres.2016.11.029

128. McCrary JM, Redding E & Altenmüller E. (2021). Performing Arts as a Health Resource? An Umbrella Review of the Health Impacts of Music and Dance Participation. PLoS ONE, 16: e0252956. https://doi.org/10.1371/journal.pone.0252956

129. Elst CFV, Foster NHD, Vuust P, Keller PE & Kringelbach ML. (2023) The Neuroscience of Dance: A Conceptual Framework and Systematic Review. Neurosci Biobehav Rev, 150: 105197. https://doi.org/10.1016/j.neubiorev.2023.105197

130. Nascimento MDM. (2021). Dance, Aging, and Neuroplasticity: An Integrative Review. Neurocase, 27: 372–381. https://doi.org/10.1080/13554794.2021.1966047

131. Balazova Z, Marecek R, Novakova L, Nemcova-Elfmarkova N, Kropacova S, Brabenec L, et al. (2021). Dance Intervention Impact on Brain Plasticity: A Randomized 6-Month fMRI Study in Non-expert Older Adults. Front Aging Neurosci, 13: 724064. https://doi.org/10.3389/fnagi.2021.724064

132. Wang Y. (2023) The Effect of Dance on Rehabilitation Training after COVID-19. Eur Rev Med Pharmacol Sci, 27: 359–365.

133. Schroeder H, Haussermann P & Fleiner T. (2023). Dance-Specific Activity in People Living with Dementia: A Conceptual Framework and Systematic Review of Its Effects on Neuropsychiatric Symptoms. J Geriatr Psychiatry Neurol, 36(3): 175–184. https://doi.org/10.1177/08919887221130268

134. Hu Y, Son K, Yang Z & Mao Y. (2023). Moderated by Personal Perception: The Preventive Relationship Between Home HIIT Dance and Depression during the COVID-19 Pandemic in China. Front Public Health, 11: 1117186. https://doi.org/10.3389/fpubh.2023.1117186

135. Humphries A, Tasnim N, Rugh R, Patrick M & Basso JC. (2023). Acutely Enhancing Affective State and Social Connection Following an Online Dance Intervention during the COVID-19 Social Isolation Crisis. BMC Psychol, 11: 13. https://doi.org/10.1186/s40359-022-01034-w

136. Raghupathy MK, Divya M & Karthikbabu S. (2021). Effects of Traditional Indian Dance on Motor Skills and Balance in Children with Down syndrome. J Mot Behav, 54: 212–221. https://doi.org/10.1080/00222895.2021.1941736

137. Kosurko A, Herron RV, Grigorovich A, Bar RJ, Kontos P, Menec V & Skinner MW. (2022). Dance Wherever You Are: The Evolution of Multimodal Delivery for Social Inclusion of Rural Older Adults. Innov Aging, 6: igab058. https://doi.org/10.1093/geroni/igab058

138. Chan JS, Wu J, Deng K & Yan JH. The Effectiveness of Dance Interventions on Cognition in Patients with Mild Cognitive Impairment: A Meta-analysis of Randomized Controlled Trials. Neurosci Biobehav Rev, 118: 80–88. https://doi.org/10.1016/j.neubiorev.2020.07.017

139. Basso JC, Satyal MK & Rugh R. (2021). Dance on the Brain: Enhancing Intra- and Inter-Brain Synchrony. Front Hum Neurosci, 14: 584312. https://doi.org/10.3389/fnhum.2020.584312

140. Wu VX, Chi Y, Lee JK, Goh HS, Chen DYM, Haugan G, et al. (2021). The Effect of Dance Interventions on Cognition, Neuroplasticity, Physical Function, Depression, and Quality of Life for Older Adults with Mild Cognitive Impairment: A Systematic Review and Meta-analysis. *Int J Nurs Stud*, 122: 104025. https://doi.org/10.1016/j.ijnurstu.2021.104025

141. Rehfeld K, Müller P, Aye N, Schmicker M, Dordevic M, Kaufmann J, et al. (2017). Dancing or Fitness Sport? The Effects of Two Training Programs on Hippocampal Plasticity and Balance Abilities in Healthy Seniors. *Front Hum Neurosci*, 11: 305–305. https://doi.org/10.3389/fnhum.2017.00305

142. Zhu Y, Zhong Q, Ji J, Ma J, Wu H, Gao Y, Ali N & Wang T. (2020. Effects of Aerobic Dance on Cognition in Older Adults with Mild Cognitive Impairment: A Systematic Review and Meta-Analysis. *J Alzheimer's Dis*, 74: 679–690. https://doi.org/10.3233/JAD-190681

143. Bojner Horwitz E, Korošec, K, &Theorell T. (2022). Can Dance and Music Make the Transition to a Sustainable Society More Feasible? Behav Sci, 12: 11. https://doi.org/10.3390/bs12010011

144. Planckaert N, Duyck W & Woumans E (2023). Is There a Cognitive Advantage in Inhibition and Switching for Bilingual Children? A Systematic Review. *Front Psychol*, 14: 1191816. https://doi.org/10.3389/fpsyg.2023. 1191816

145. Korenar M, Daller JT & Pliatsikas C. (2023). Dynamic Effects of Bilingualism on Brain Structure Map onto General Principles of Experience-Based Neuroplasticity. *Sci Rep*, 13: 3428. https://doi.org/10.1038/ s41598-023-30326-3

146. Korpinen E, Slama S, Rosenqvist J & Haavisto A. (2023). WPPSI-IV and NEPSY-II Performance in Mono- and Bilingual 5–6-Year-Old Children: Findings from The FinSwed Study. *Scandinavian J Psychol*, 64: 409–420. https://doi.org/10.1111/sjop.12895

147. Huang R, Baker ER & Schneider JM. (2023). Executive Function Skills Account for a Bilingual Advantage in English Novel Word Learning among Low-Income Preschoolers. *J Exper Child Psychol*, 235: 105714. https://doi. org/10.1016/j.jecp.2023.105714

148. Calvo N, Anderson JAE, Berkes M, Freedman M, Craik FIM & Bialystok E. (2023). Gray Matter Volume as Evidence for Cognitive Reserve in Bilinguals with Mild Cognitive Impairment. *Alzheimer Dis Assoc Disord*, 37: 7–12.

149. Ovando-Tellez M, Kenett YN, Benedek M, Bernard M, Belo J, Beranger B, et al. (2022). Brain Connectivity–Based Prediction of Real-Life Creativity Is Mediated by Semantic Memory Structure. *Sci Adv*, 8: eabl4294.

150. Ahmed H, Pauly-Takacs K & Abraham A. (2023) Evaluating the Effects of Episodic and Semantic Memory Induction Procedures on Divergent Thinking in Younger and Older Adults. *PLoS ONE*, 18(6): e0286305. https://doi. org/10.1371/journal.pone.0286305

151. Ramscar M, Hendrix P, Shaoul C, Milin P & Baayen H. (2014). The Myth of Cognitive Decline: Non-Linear Dynamics of Lifelong Learning. *Top Cogn Sci*: 1–38. ISSN:1756–8757 print / 1756–8765 online. https://doi.org/10.1111/ tops.12078

152. John Kander, Lin-Manuel Miranda on "New York, New York" on Broadway - CBS News (Accessed online October 7, 2023).

153. Edwards EE, St Hillaire-Clarke C, Frankowski DS, Finkelstein F, Cheever T, Chen WG, et al. (2023). NIH Music-Based Intervention Toolkit Music-Based Interventions for Brain Disorders of Aging. *Neurology*, 100: 868–878. https://doi.org/10.1212/WNL.0000000000206797

154. Tanzi R. Neuroscientist & Musician Performing at nlm.com/rudytanzi

155. Jessberger S & Gage FH. (2008). Stem-Cell-Associated Structural and Functional Plasticity in the Aging Hippocampus. *Psychol Aging*, 23(4): 684–691.

7

DIET

Reduce Inflammation and Nourish Your Building Brain

The U.S. Department of Agriculture and U.S. Department of Health and Human Services. Dietary Guidelines for Americans, 2020–2025. 9th Edition, available at https://www.dietaryguidelines.gov/sites/default/files/2020-12/ Dietary_Guidelines_for_Americans_2020-2025.pdf, emphasizes the profound impact of foods and fluids consumed on overall health across the lifespan. It notes that "core elements of a healthy dietary pattern are remarkably consistent across the lifespan and across health outcomes." While knowing and observing calorie limits, foods and beverages that are nutrient-dense are recommended because they provide minerals, vitamins, and other molecules that promote health without added sugars.

Adequate hydration is also essential for good cognitive functioning.[1] Individuals with dementia were found to have more severe cognitive deficits when dehydrated.[2] Research is essential "to define clearly if dehydration is a cause or effect of the onset of dementia." Figure 7.1 invites us to enjoy healthy intake.

On the farm of my youth the amount and variety of foods and fluids on the farm of my youth were copious. Whole grains, nuts, vegetables, and fruits? Grew them. Harvested them. Kept them in the pantry.

However, much of what we consumed was less than wise. A common snack was a slice of homemade white bread with a thick layer of home-churned butter into which was pressed a large amount of white sugar.

Hopefully, your oral intake is healthier, because these selections could induce neuroinflammation; glycation which creates advanced glycation end

DOI: 10.4324/9781003462354-7

Figure 7.1
This illustration by David Yu, MD, of a brain cell serving some healthy foods highlights the role of diet in reducing neuroinflammation and nourishing your building brain.

products (AGEs) that are key factors in aging and diseases such as Alzheimer's; and other bad side effects. Evolving neuroscience provides data to influence intake of foods and fluids that can reduce inflammation and nourish your neurons to build bigger, better brains.

Choosing foods wisely includes no added sugar. In a recent, lengthy, information-packed article, Lustig[3] suggested that sugar needs to be regulated. I agree with the content and perspective of that article and would add that using stevia in lieu of sugar could eliminate the detrimental impact of sugar, assist food producers in creating tasty foods for lower cost, and might include advertising that these delicacies provide health benefits described herein. Educating the food industry to the various ways that stevia could be beneficial to their bottom line and to their consumers healthy longevity could hasten the shift of the public to healthier foods. More on this later in this chapter.

The Dietary Guidelines indicate that core elements of a healthy diet include all types of vegetables; whole fruits; whole grains; dairy products and/or soy beverages; protein from lean meats, seafood, eggs, peas, beans, nuts, seeds, lentils, and soy products; and oils in nuts, seafood, and other foods. The USDA makes a personalized food plan available at https://www.myplate.gov/. The fast-food egg, cheese, bacon, high fat, and high salt breakfast available at the drive-through has already been given extensive attention as bad for long-term health. More recently research found 15–20% decreased blood flow within two hours of eating this type of meal.

Maximizing brain chemistry, architecture, and performance *AT ANY AGE* requires good nutrition across the lifespan. Ideally a healthy diet starts as early as in utero with proper nutrients during the pregnancy.

The World Health Organization (WHO) recommends breastfeeding as the exclusive source of nutrients during the first six months of life with breastfeeding continuing for at least the first two years of life because "breastfeeding fosters healthy growth, improves cognitive development, and may have longer term health benefits."[4] For example, healthy growth has included increased thickness in the brain cortex as well as higher Full Scall IQ in adolescents who were breastfed.[5]

We can celebrate that WHO set a "global target to increase exclusive breastfeeding by 2025." WHO Director General and the Executive Director of UNICEF noted significant progress in the recent ten years of exclusive breastfeeding and urged that we "improve progress on breastfeeding rates globally," as found on the WHO website 3 March 2024.[6]

WHO also points out that the

> exact make-up of a diversified, balanced, and healthy diet will vary depending on individual characteristic (e.g., age gender, lifestyle, and degree of physical activity), cultural context, locally available foods, and dietary customs. However, the basic principles of what constitutes a healthy diet remain the same.

Dietary impact on telomeres is one of the ways to influence overall health and longevity. Telomeres are the proteins that protect the end of chromosomes and have important functions in how DNA is copied correctly to preserve your genetic information as early as in utero. Maintaining the length of telomeres is associated with better health throughout a vigorous longevity. Maintaining a healthy diet pattern for good metabolic status can protect the telomere length[7] with increased protection when including some micronutrients as will be discussed below. Oral intake that reduces inflammation and oxidative stress or influences epigenetic reactions could help maintain the length of telomeres. As factors associated with maintaining the healthy length of telomeres are addressed herein, that good telomere influence will be accentuated herein.

Okinawan research across time most succinctly captures the essence of what we need to learn, know, and live by.[8] Please note that the World Health Organization had, at one point, considered Okinawa to be one of the Centenarian Centers of the World. It is unfortunate that the early decades of that research have been underappreciated because it was written in Japanese. It is indeed fortunate that we now have sufficient information from that database to capture the wisdom of their elders *and* fortunate that evolving neuroscience enriches that database. In this chapter the focus is on aspects of the Okinawan diet that were part of the trajectory to being recognized as a centenarian center of the world; we will also visit the influences of Western cultures that led to loss of that distinction. This included the Western diet consumption of too many calories which can accelerate the decline in neurogenesis and increase in systemic, sterile, low-grade inflammation (inflammaging) that are noted in unsuccessful aging and neurocognitive decline. Less intake of foods high in nutrients and greater intake of unhealthy "Western" foods were each associated independently with less volume in the left hippocampus.[9]

Recent clinical and in vitro research increased data on how gut organisms influence production of metabolites which can protect cognitive reserve by reducing neuroinflammation.[10] Culturing human astrocytes with this metabolite resulted in cellular changes supporting the role of the gut microbiome in healthy brain aging. Comparing fecal samples of people 60 years old and older with MCI ($n = 23$) as compared with the same-aged cognitive healthy group found a microbiome that was less diverse in subjects with MCI.[11] Dietary patterns such as the Mediterranean diet[12] and Okinawan diet could have a healthy influence on the microbiome. In vitro studies suggest that consuming stevia could have a positive influence on the gut microbiome; studies are needed to assess frequency, amount, with which foods, and timing of consumption that could maximize this benefit.[13] Other proposed interventions to promote healthy microbiota for brain health include high-fiber diets and physical exercise.[14]

Fat cells can increase inflammation which can influence much of your body. Inflammation is bad for brain function and is a risk factor for many health problems including dementia. This accentuates the Okinawan wisdom to *hara hachi bu*, "eat until you are 80% full" of healthy calories.

A subset of the Women's Health Initiative which focused on memory found that reducing fat intake 20% and increasing fruit, vegetable, and grain intake was associated with "significantly reduced risk of possible cognitive impairment in postmenopausal women."[15] Studies suggest that diet patterns can influence cognitive function and that the strongest benefits are realized with earlier use of the healthy choices; as mentioned above and recommended by WHO, that includes during pregnancy and breast feeding to maximize positive influences on brain development from the very beginning of life.

With the Okinawan diet you will likely get adequate vegetables including fruits, sweet potatoes, fish as a source of animal protein and healthy fats, and little or no sugar. Five portions per day of fruits and of vegetables is considered adequate. That can easily be achieved by choosing from a variety of fruits and vegetables; by including a vegetable with each meal; by eating fruits and vegetables for snacks; and by eating fresh sources that are in season. Fat intake and fat sources on the Okinawan diet are likely to be healthier because they come from fish as the primary source of animal protein.

The Mediterranean diet is similarly rich in vegetables and fruits. Autopsies on 581 participants in the Rush Memory and Aging Project found associations between lower global Alzheimer's pathology and dietary patterns.[16] Those who had been on the Mediterranean diet had the lowest beta-amyloid load independent of vascular disease, smoking, and physical activity. "Those in the highest tertile of green leafy vegetables intake had less global AD pathology when compared with those in the lowest tertile." Their average age at death was 91 years.

A recent study found better value in the Mediterranean diet when using extra virgin olive oil rather than nuts. Perhaps that is related to the high concentration of oleic acid, a monounsaturated fatty acid in olive oil and in avocados. Oleic acid promoted survival of neural stem cells in an in vitro study using "purified human neuroprogenitors derived from human embryonic stem cells."[17] Neural stem cells in our adult hippocampus can mature into newborn brain cells that are essential for memory, new learning, mood regulation, and stress management.

Changing your eating behaviors, rather than focusing on any single dietary component, might improve your cognition and might decrease your risk of and/or delay dementia onset. Both the Okinawan and the Mediterranean diets have been associated with better cognitive performance and reduced risk of dementia. This might add up to several more years of celebrating senior successes. That is, in part, related to the synergistic effect of the many bioactive components of foods.

There is a possibility that the Mediterranean diet might be associated with better maintenance of telomeres, the protective caps at the end of DNA strands. It is likely that by eating either the Okinawan diet or the Mediterranean diet you could also enjoy the antioxidant and anti-inflammatory benefits of getting sufficient polyphenols; the best source of carbohydrates and fiber; no added sugar; animal and vegetable proteins; and the right kinds of healthy fats.

We will focus on the Okinawan diet, Mediterranean diet and other protocols that fit what is shown in evolving neuroscience. In the spirit of Positive Psychology[18] we will start with getting the difficult aspects covered so we can immerse ourselves in the rich and delicious positive research.

Difficult?

To Eat or Not to Eat ... That Is the Question

Neuroinflammation is thought to be a major factor in losing brain health and brain power. Also difficult for some of us? One of the most powerful ways of avoiding and/or reversing neuroinflammation is calorie restriction.

It is significant that part of the Okinawan lifestyle included *hara hachi bu*, translated as "leave the table when you are 80% full." Research suggests that *hara hachi bu* or "eat until you are 80% full," had been an important factor in exceptional longevity with increased healthspan for the humans in Okinawa in research that has been ongoing since 1975. In that research and elsewhere, it has been noted that centenarians age more slowly and delay, even escape, chronic diseases associated with aging including cardiovascular disease, cancer, and dementia. Li and colleagues[19] found "uniquely enriched gut microbes" in a small population of centenarians which, they opined, might promote health and extreme longevity.

Calorie Restriction

Reducing caloric intake has also been associated with improved learning and memory as well as improving the resistance of the brain to various forms of damage according to a review by Fusco and Pani.[20] With calorie restriction (CR) and good nutrition, animals remain "younger longer" with less disease, less accumulation of body fat, better regulation of insulin and glucose, and continued good functioning of mitochondria (little energy factories in most cells). How this relates in humans awaits data from ongoing research in quest of more clarity.

Autophagy might become a new favorite four-syllable word. Let's explore why.

Autophagy is part of the human body's housekeeping system. It is a bulk recycling general process; we might think of it as refurbishing your cells in that it includes degrading and removing damaged or dysfunctional cell components. Autophagy can be increased with influences such as exercise or calorie restriction because the nutrient and energy levels signal this process and can influence health and lifespan.

This research influenced me to do my exercise routine prior to any caloric intake. Roll out of bed, go to the bathroom, drink a glass of water with my multivitamin, get on the exercycle with neuroplasticity research to read (or meditate on), and wake up while aerobic. Fun!

By consuming less calories than are needed for exercise, this increases the autophagy process of using the fat molecules in existing fat cells for the essential energy to accomplish the physical activities. When exercise and decreased calorie intake deplete the liver glycogen stores, fatty acids in adipose tissues are converted to ketones.[21] Recent research suggests that fasting increases the ketone β-hydroxybutyrate which can increase BDNF which can protect mitochondria, promote plasticity at the synapses, and influence better management of stress at the cellular level.[21]

This is one of the reasons that a person's body fat percentage is more important than their weight. Watching mine decline has been encouraging as well as empowering.

Meanwhile, staying active and eating wisely are among the many efforts essential to improve brain health, neurocognitive functioning, and general health. Neither calorie restriction nor intermittent fasting should be taken lightly. Both will require meeting basic dietary requirements.

The efficiency of the human body's elite and complex housekeeping/bulk recycling process of autophagy might decline with age. However, autophagy remains essential because it protects cells, including brain cells.

Increasing autophagy might enhance the activities of cells in the hippocampus, the part of the brain most associated with memory and learning. Some researchers have shown that restoring the level of autophagy in the hippocampus promoted better formation of memories and integration of information[22] These researchers say that restoring autophagy to a more youthful level was "sufficient to reverse age-related memory deficits" and other cognitive impairments associated with advanced ages.

Calorie restriction (CR) that includes good nutrition has shown significant cognitive benefits in humans and in many lower animals. Researchers[23] opined that "the take-home message from CR research is that the rate of aging can be manipulated. Factors that contribute to age-associated diseases can be influenced to disconnect chronological age from biological age." Perhaps that makes autophagy one of the best housekeeping and recycling systems on the planet.

Intermittent Fasting

The current level of neuroscience on intermittent fasting and/or calorie restriction might make them more enticing. There are several definitions of intermittent fasting.

The status of research suggests that time-restricted eating influences metabolic changes that may improve function of mitochondria for energy production, enhance plasticity at synapses, increase autophagy, increase insulin sensitivity, and thereby improve cerebrovascular functioning, and improve neurocognitive functioning.[24] As ongoing research in this realm continues, details on the potential neuroprotective effects of time-restricted eating will evolve.

This could be effective in reducing inflammation, protecting telomeres, increasing autophagy, and having other benefits. A research review[24] suggests that this lifestyle change may be neuroprotective by reducing neuroinflammation so that cognitive skills can be retained.

Two Days a week
Perhaps the most difficult example is total fasting for two days of each week. For me, this has *no* appeal. There are more easily accomplished forms of intermittent fasting.

Sixteen to Eighteen Hours Daily of Zero Calories
With the absence of calorie intake for 16–18 hours, autophagy is likely to increase, and metabolism shifts to burning adipose tissue. For most people, that could be a healthy outcome. Autophagy might be activated in the late portion of some periods of fasting and could increase the number of stem cells to enhance regeneration in some body tissues when feeding is resumed. This could be followed by "improved metabolic function, reduced inflammation with delayed immunosenescence, reduced oxidative damage, and improved proteostasis."[25] This could delay and/or reduce risk factors for and incidence of neurodegenerative diseases.

Some researchers say that restricting your calorie intake to six or eight hours each day could increase autophagy, neurogenesis, and healthspan for a vigorous longevity.[26] Fasting increases ketone bodies which have "profound effects on systemic metabolism." These ketone bodies "stimulate expression of the gene for brain-derived neurotrophic factor" (BDNF) which improves brain health and decreases risk for "psychiatric and neurodegenerative disorders." Intermittent fasting influences cells to respond to stress and increase antioxidant defenses. Since both autophagy and mitophagy are stimulated, cells are enabled to "remove oxidatively damaged proteins and mitochondria and recycle undamaged molecular constituents while temporarily reducing global protein synthesis to conserve energy and molecular resources." These impacts of fasting can be enhanced with exercising on a regular basis to improve physical and mental performance as well as resistance to disease and neurologic disorders.

My established pattern is consuming all calories between 7 AM and 3 PM and completing an exercise routine before 7 AM prior to any calorie intake because autophagy is signaled by the nutrient/energy level. This is only one example of one person's attempt to "live off the fat of the land," and even though this reduced my body fat percentage across time with this protocol, this is a prompt to emphasize the importance of you conferring with your healthcare professionals to personalize a strategy to maximize the many benefits of calorie restriction including time-restricted intake.

This is also an example of how artificial intelligence could hack neuroplasticity by capturing the brain activity during exercise, the time when calories are consumed, calories used during exercise, and the changes in brain activity across the hours of having no calories. Given the obesity crisis in many citizens, this is an urgent call to develop this artificial intelligence aid in our Intelligent Neuroplasticity Revolution (INR).

A review of clinical and preclinical studies found vasoprotective benefits of fasting[27]; these are protective of neurocognitive functions. Clinical studies of intermittent fasting have found benefits for several health conditions, including diabetes, hypertension, obesity, cancers, cardiovascular disease, and neurological disorders.

Ramadan

Ramadan fasting of healthy adults during daylight hours for one month is the most frequently practiced intermittent fasting. In a small study with 34 men aged 16–64, Ramadan fasting was associated with improved lipid panels, lower blood pressure, less inflammation and oxidative stress; it was also associated with significant decrease in fat intake, calories consumed, waist circumference and weight.[28] Further research with more participants and including women is needed to clarify these potential health benefits.

To Sum It Up

Total intake of food and fluid, frequency of intake, and content consumed all factor into the molecular events of energy metabolism and neuroplasticity.[29] Nutrient molecules influence the health and function of a neuron, how it transmits signals, and the plastic response at the synapse. The optimal combination of nutrients can be a practical way of enhancing cognitive performance while increasing the health span. Full coverage of all pertinent research is not attempted in this writing; rather the intent of this writing is to highlight several dietary choices that could be neuroprotective,[30] could have positive effects on neuroplasticity[31] including on adult neurogenesis,[32] and could influence the reduction of chronic inflammation[33] which has deleterious effects on brain health and brain function.

Nutrition across the lifespan can include epigenetic modifications.[34] This might include "potential reversal of negative epigenetic marks." More research is needed in human subjects to further define how nutrients affect epigenetic influence on brain performance and development across the lifespan.

A recent review of many studies provides an overview on nutrition for improving healthy aging of cardiovascular and brain tissues to prevent age-related diseases.[35] Studies reviewed "show that the development of cardiovascular and cerebrovascular diseases, neurodegenerative diseases, cognitive impairment and dementia can be slowed down or prevented by certain diets with anti-aging action."

Brain Building Basics with Macronutrients

As ongoing research in this realm continues, details on the potential neuroprotective effects of time-restricted eating will evolve. In addition to time, other basic and essential needs are the number of calories and how these are obtained from carbohydrates (CHO), fats, proteins, and fiber.

Modifying intake to reduce weight requires special considerations. For example, maintaining bone mineral density is essential for successful aging. A review of the research indicates that proper physical training while getting adequate nutrition could include increasing bone mineral density in the hip and femoral neck during weight reduction. This is pertinent to protecting your brain power *at all ages* since it could reduce the risk of injuries from falls.

Given the opinions of many scientists, it is very important that about one third of dementia cases can be attributed to modifiable risk factors which includes eating for health. Current dietary recommendations, such as those in the WHO Guidelines, are likely to enhance and maintain good brain power *at any age* while reducing the incidence and prevalence of dementia. The same could be said for the recently released Dietary Guidelines for Americans.

However, I agree with the 20 academics and doctors on the scientific committee that recommended the limit for added sugars in our diet be reduced from 10% to 6% of daily calories. They made this recommendation in response to the rapidly rising rate of obesity and its link to many health problems including cancer, heart disease, and Type 2 diabetes.

For these and similar reasons I would go even further and recommend that there be *no* added sugars in our diet so that we can avoid the health problems associated with advanced glycation end products (AGEs) which include neurodegeneration. As described herein, sweetening our food with stevia adds no calories, does not advance the aging process through accumulation of AGEs, and might give us several significant health benefits such as reduced waist circumference and weight loss.[36]

This rich and delicious topic of eating for general health and brain health has sufficient neuroscience evidence available to fill a very large book that would be too heavy to carry, too lengthy to read. Thus, the goal in this writing is to shed light on most of those dietary elements which research suggests could be beneficial in several ways: build a healthy brain; be neuroprotective; could enhance how brain cells change with experience; might increase the creation of new healthy brain cells; and could influence the reduction of inflammation which has been associated with the development of some serious brain disorders.

Our Diet to Reduce Neuroinflammation and Nourish Your Building Brain begins with the basic tenet that energy in needs to approximate energy out. It emphasizes that every molecule taken in should be nutritious and have minimal, preferably no negative effects.

Conferring with a healthcare provider on how to use this evolving neuroscience in designing a uniquely personalized best diet for general health and brain health is important throughout the lifespan. We can celebrate how using artificial intelligence can help us hasten the accuracy and use of evolving neuroscience which has become available close to real time. Working together we can also influence a multitude of ways that artificial intelligence can influence how we personalize and maximize our unique Intelligent Neuroplasticity Revolution.

Calories

Resting energy expenditure (REE) is the energy expenditure of an individual who is not fasting, and it is the number of calories required for a 24-hour period by your body during a non-active period. REE usually accounts for more than 60% of the total energy expenditure.[37]

Carbohydrates (CHO)

It has been suggested that unrefined, complex carbohydrates should be 45–60% of calorie intake daily[25] with sources from vegetables, fruits and complex grains. Cognitive test data were obtained from 2,537 individuals 60 years old and older who participated in the NHANES study. Results suggested that a low carbohydrate diet might be related to decreased cognitive performance.[38] A strength of this study is the large number of participants providing data. However, it is significant that this data was obtained by two interviews rather than direct observation. It also required individuals to recall their food intake for a 24-hour period. Thus, further studies are needed to clarify the potential for low carbohydrate intake to decrease cognitive performance.

Artificial intelligence could facilitate this data acquisition for ongoing essential research. Artificial intelligence could also personalize prompts for identified evidence-based behaviors; record these activities; and provide personalized positive reinforcement for an individual's path to the best CHO intake for enhancing neuroplasticity, brain health, and neurocognition.

Fats

Calorie intake and energy expenditure should balance to avoid increasing adipose tissue. The WHO Healthy Diet Fact Sheet[39] suggests that a maximum of 30% of caloric intake be from fat. Of this, less than 10% should be saturated fats and trans-fats in total oral intake should be less than 1% with "the goal of eliminating industrially produced trans fats." They emphasize that "Unhealthy diet and lack of physical activity are leading global risks to health." This can be accessed online.[39]

Crucial to optimal central nervous system structure and function are the essential omega-3 fatty acids eicosapentaenoic acid (EPA) and docosahexaenoic acid (DHA).[40] Primary sources include fish and plant foods. Interestingly, in the parts of the brain essential to cognition and memory one study found increased gray matter volumes that were associated with fish consumption and independent of plasma measures of omega-3 fatty acids,[41] highlighting the complexity of, and interactions of, dietary impact in humans. In the Framingham Heart Study people whose DHA level was in the top quartile had a highly significant 47% lower risk for developing dementia.[42]

The essential omega-3 docosahelaenoic acid (DHA) is critical for brain development and function from conception throughout the lifespan.[43] Eicosapentaenoic acid (EPA) is another essential omega-3 fatty acid. Essential omega-3 fatty acids are so called because our bodies do not create them. However, our central nervous system must have them for both structure and function. Approximately 25% of brain cell membranes consist of these fatty acids. They effect the growth of neurons, changes in neuron membranes, gene expression, and signal transduction. They protect against neurodegenerative pathologies, reduce neuroinflammation, protect functions of mitochondria, and reduce oxidative stress. They are essential "during the first year of life and in the developing/prevention of neurodegenerative diseases associated with aging."

A random controlled trial[44] that supplemented with fish oil found "increased red blood cell omega3 content, working memory performance, and BOLD signal in the posterior cingulate cortex during greater working memory load in older adults with subjective memory impairment." Thus, they suggested that supplementing with omega-3 fish oil could enhance brain cell response to challenges in working memory.

The best sources of DHA and EPA in our diet will be fatty fish such as mackerel, sardines, tuna, and salmon. Animals fed a diet high in omega-3 fatty acids early in life had less damage from environmental challenges later.[29] When they were transitioned to a Western diet, their "epigenetic memory" protected them from loss of brain performance. These researchers believe that the diet high in omega-3 fatty acids affected their genetic function in ways that kept them less vulnerable to the negative impact of the Western diet later.

For me, a tin of sardines with a few wholegrain crackers is a meal that is delicious, healthy, and easy to take to any place of employment. However, none of my coworkers have ever accepted my offer to share this!!

These omega-3s are also needed for building and maintaining the myelin sheath on your brain cell axon, the branch leaving the cell body that goes to the appropriate location carrying information. The health of the myelin sheath is essential for efficient exchange of information between cells as well as for brain cell health. Working memory in healthy humans improved significantly after six months of ingesting omega-3s that contained 750 mg per day of DHA and 930 mg per day of eicosapentaenoic acid (EPA), two essential omega-3s. DHA in the brain cell membrane helps ease transmission of molecules across this membrane to get information between brain cells to maximize learning, memory, and sleep.

Among community-dwelling individuals with mild cognitive impairment, only those with good omega-3 status showed benefit when given vitamin B supplements. They had 30% less brain atrophy and their memory and global cognition were maintained better.

This illustrates that these various nutrients form a complex and interactive system. Satizabal and colleagues[45] found that higher omega-3 levels in human red blood cells was associated with larger volumes of the hippocampus as well as better abstract reasoning even in cognitively healthy adults. Thus, omega-3 intake might promote cognitive resilience.

Proteins

Building and maintaining adequate muscle across vigorous longevity requires both exercise and adequate dietary protein. Your healthcare provider can help determine the ideal amount of protein in your diet. Some researchers suggest a daily intake of 1.2 grams of protein per kilogram of body weight. Individuals on the form of calorie restriction that limited their eating hours to 7.5 hours per day consumed 1.6–1.9 grams per kilogram of body weight each day throughout the research protocol; those with the time restriction had comparable changes in lean mass and skeletal muscle changes as compared to the control group that ate the same amount of protein freely through the day.

Since timing of protein intake is important for best muscle building associated with exercise, it is after my aerobics and strengthening exercise routine in the early morning that I consume proteins. My favorite is the combination of unflavored protein powder and stevia mixed with cocoa powder for the flavonoids. Every molecule in it is a molecule of health. If I was sweating a lot, there might be a few grains of salt added.

The source of the protein consumed could be important.[46] Comparing the highest quintile of consumption with the lowest quintile when the source of the protein consumed was from plants, there was an inverse relationship between amount consumed and death from dementia, death from cardiovascular disease, and from all-cause mortality. They found higher risk of death from dementia with consuming processed red meat whereas consuming poultry and eggs was associated with a lower risk of death from dementia.

This prospective cohort study "included 102 521 postmenopausal women enrolled in the Women's Health Initiative between 1993 and 1998" who were followed "through February 2017."

Leucine is one of the important amino acids in proteins for stimulating muscle growth. Good food sources of this important protein include, but are not limited to, canned navy beans, chickpeas, salmon, and soybeans. Tofu is one of my favorite sources of this important protein.

The amount of protein consumed probably needs to be higher in our oldest old (people 85 years old and older) because the ways that proteins are digested and utilized changes across the lifespan. Please note that, since the metabolic outcome of protein digestions includes production of large molecules, it is essential to get adequate water to rinse these through your kidneys. At least 8 glasses of water per day is a usual recommendation.

Spermidine is an example of the importance of protein intake because of the antiglycation effect.[47] As described above, glycation is a process of proteins and other molecules in your body reacting to sugar molecules resulting in decreased protein function and reduced elasticity in your skin, tendons, and blood vessels through irreversible rearrangements leading to creating advanced glycation end products (AGEs).

Can we agree that this is one use of the word AGEs that we *all* want to avoid?

Spermidine sources include, but may not be limited to, wheat germ, soybeans, corn, green peas, and legumes. Tofu is a favorite source of mine. Pomegranate is a fruit that may also help reduce glycation.[48]

Fiber

Several fruits can be beneficial for brain health because of their many nutrients and fiber which can influence gut microbes to metabolize polyphenols. Diets high in fiber promote growth of bacteria that aid in digestion.[49]

A review looked at 61 studies which focused mainly on depression but included some other components of quality of life, flourishing, and mental well-being.[50] Findings suggested that high vegetable and fruit intake "may promote higher levels of optimism and self-efficacy." These positive mental health outcomes, although found with various populations and using different research design which are limitations to applicability, suggest possible benefit for mental health in consuming "at least 5 portions of fruit and vegetables a day." Although more research is needed, vegetable and fruit positive influences on mental health are multifaceted.

Researchers considered data from 2,478 participants in the NHANES study.[51] Their findings led them to opine that consuming sufficient fiber could be preventive of cognitive impairment in elders with uncontrolled hypertension.

Micronutrient Matters

Vitamins and Minerals

Improved cognition was found in elders given a commercial multivitamin-mineral daily.[52,53] There were 2,262 participants in this three-year study. They had "improved global cognition, episodic memory, and executive function." The tests included

> the 50-point modified Telephone Interview for Cognitive Status (TICSm with 10-minute short delay word list recall), an additional 40-minute Long Delay Word List Recall, immediate and delayed Story Recall (SRI & II), Oral Trail-Making Test Part B (OTMT-B, log transformed; Part A was administered for practice only), Verbal Fluency by category (VF-C) and letter (VF-L), Number Span (NS), and Digit Ordering Test (DOT).

Part of the brilliance of this study was having a control group that was given a low dose of cocoa flavanols. The finding of no cognitive benefit with cocoa extract may be related to the low dose of 500 mg. which research associated with arterial stiffness whereas, as described below, it was 900 mg per day of cocoa flavanols for three months that was associated with better brain performance on cognitive testing as well as on fMRI.

A review of research on the association of vitamin B12 with Alzheimer's disease[54] found that "cell culture and ex vivo studies provided further evidence for the protective effects of vitamin B12. These are linked to amyloid formation and fibrillization, epigenetic modifications, tau fibrillization, synaptogenesis of neuronal membranes, oxidative stress, and cholesterol synthesis." Vitamin B12 combined with other B vitamins improved cognitive function, reduced inflammation, and reduced brain atrophy in healthy elders as well as in those with mild cognitive decline. Lower plasma levels of vitamin B12 were found in patients with Alzheimer's disease as compared to healthy elders. Improved cognitive functioning was noted with B vitamin supplementation. Since animal-based food is a good source of vitamin B12, vegetarians and vegans might be advised to monitor their B12 status to ensure they get sufficient antioxidant and anti-inflammatory benefits of this neuroprotective nutrient.

Other vitamins of note are the "neurotropic B vitamins thiamine (B1), pyridoxine (B6), and cobalamin (B12)" which are neuroprotective.[55] "While vitamin B1 acts as a site-directed antioxidant, vitamin B6 balances nerve metabolism, and vitamin B12 maintains myelin sheaths." They support new cell structural development.

The VITACOG trial included 266 people at or above 70 years old who had mild cognitive impairment.[56] Participants who received high daily doses of vitamin B12, B6 and folic acid for two years were assessed for omega-3 fatty acids at baseline as well as with serial MRI scan evaluations of whole brain changes in rate of atrophy. Individuals receiving the B vitamins for two years had "final scores for the verbal delayed recall, global cognition and clinical dementia rating" that were related to omega-3 baseline concentrations. The higher docosahexaenoic acid concentrations alone "significantly enhanced the beneficial cognitive effects of B vitamins." These researchers opined that "a combined supplementation of B vitamins and omega-3 fatty acids is suggested as potential therapy to slow the conversion from MCI to AD" and they recommended further studies to investigate this. When the baseline omega-3 was high at baseline, subjects receiving the B vitamins had 40% less brain atrophy.[57]

A review of random controlled trials on dietary supplements[58] found that "omega-3 fatty acids such as DHA and eicosapentaenoic acid (EPA) seem to be the most promising nutritional intervention in the prevention of AD." Uptake of carotenoids, which are hydrophobic, was increased when taken with these fatty acids resulting in improved mood, memory, and sight. Larger interventional trials of this combination of vitamins and omega-3 fatty acids to assess neurocognitive benefits are needed.

Optimal nutrient combinations to enhance brain health and prevent disease are not yet determined. The interactions between foods and nutrients are complex. Thus, diet patterns have been investigated. For example, combining ginger and avocado might be useful in working with obesity. A recent review of research on this combination found improved blood lipids as well as antioxidant and anti-inflammatory impact with no negative effects.[59] Components of ginger reduce neuroinflammation.[60] As compared to adults not consuming avocado, adults consuming avocado had significantly better "overall global cognition score, which remained significant when controlling for all relevant confounders."[61] In an in vitro study, "gingerol significantly ameliorated neuronal damage" and increased

BDNF levels.[62] These researchers noted that "BDNF is a critical contributor to neuronal processes and has emerged as a key therapeutic target in neurodegenerative diseases."

Polyphenols

Since DHA is vulnerable to oxidative damage, polyphenols are essential. Polyphenols are molecules produced by plants as a protection against ultraviolet radiation and pathogens. Polyphenol benefits include anti-inflammatory effects, antioxidant effects, antiaging impact, reduction in beta-amyloid deposits, autophagy activation, and increasing neurogenesis as well as the health and survival of the new brain cells.

Polyphenols affect diversity of gut microbes that favors inhibition of pathogenic bacteria and increasing beneficial bacteria.[63] This is important in promoting a healthy gut-brain axis. Polyphenols also have antioxidant and anti-inflammatory effects. Since polyphenols can cross the blood-brain barrier, their anti-inflammatory and antioxidant effects are neuroprotective. They may also have "neurogenic effects"[64] although further research tools and protocols must be developed before this impact on increasing neurogenesis can be assessed in humans.

Humans get polyphenols by eating fruits, vegetables, spices, grains, and other plant-based foods. An important polyphenol example is flavonoids because of their positive cognitive and anti-inflammatory effects. Much of this research was done with other animals; however, scientific advances have developed a unique tool. Scientists can now cause induced pluripotent stem cells to generate nerve and glial cells in their labs so that in vitro studies can add to our understanding of flavonoid metabolites bioavailability and actions in humans and dramatically increase the relevance of other studies.

Food frequency and cognitive performance were assessed annually across 6.9 years with 961 Rush Memory and Aging Project participants whose ages were 60–100 years.[65] "Higher dietary intake of total flavonols and flavonol constituents were associated with a slower rate of decline in global cognition and multiple cognitive domains." Cognitive domains with slower decline included working memory, perceptual speed, and semantic memory. These researchers opined that "dietary intakes of total flavonols and several flavonol constituents may be associated with slower decline in global cognition and multiple cognitive abilities with older age." Apples, kale, beans, spinach, broccoli, tomatoes, pears, olive oil, and oranges were among the food sources of flavonoids for these 961 subjects who demonstrated more years with continuing global cognition, perceptual speed, semantic memory and working memory.

Cocoa

This source of polyphenols is my favorite!! Mixed with a flavorless protein powder, stevia, and sufficient water to create a pudding consistency, it has been a party favorite that people ask for more of. No guilt with this sweet pudding. Every molecule in it is a molecule of health.

The flavonoids found in cocoa are noted for powerful anti-inflammatory as well as antioxidant effects. Another benefit of these flavonoids is the improved blood flow to the brain as well as increased health and flexibility of blood vessels.[66,67] The positive effect of acute as well as of chronic consumption of cocoa on blood vessel endothelium might be dose-dependent in that more increase in flow-mediated dilation of the brachial artery has been noted two hours after participants consumed larger amounts of cocoa.[66] Although more research is needed to establish the most favorable intake for flow-mediated dilation and other benefits, as little as a teaspoon of natural cocoa powder might suffice.

Does not sound like a hardship to me. Might be a favorite formula for prevention.

A Japanese group of young students had four weeks of consumption of 508 mg cacao polyphenol each day during a study that included exercising at 50% of maximal oxygen uptake.[68] This high intake of cocoa during intense exercise for four weeks reduced the stiffness of their arteries.

Vascular endothelial health and flexibility are very important because decreased cerebral blood flow and poor regulation of blood flow to the brain are likely to result in cognitive dysfunction and dementia in Alzheimer's disease as well as in vascular dementia. Your brain is a very high utilizer of oxygen and glucose. Both the structural and functional integrity of your brain requires perfect regulation of the blood supply matched to its need for the energy supplied by these nutrients. Several cardiovascular risk factors cause damage and stiffness to the walls of your arteries; these factors include but might not be limited to high LDL cholesterol, high blood pressure, diabetes or high blood sugar, inflammation, obesity or being overweight, and physical inactivity.

Any damage and stiffness in blood vessels could also impair the drainage system of your brain; thus, less beta-amyloid is cleared from the fluid surrounding and protecting your brain and more can accumulate in the plaques associated with developing Alzheimer's disease. Increasing the health and flexibility of your blood vessels, thus, is doubly important in that removing toxins is as important as supplying adequate nutrients to maximize brain health and performance. Improvements found in cognition after ingesting cocoa might be related to better control of some of the cardiovascular risk factors such as lower blood pressure; improved health and function of linings of your blood vessels; improved sensitivity to insulin with improved glucose metabolism; and healthier cholesterol.

Healthy elders might reverse these risk factors in two to four weeks by consuming 450 mg of cocoa flavanols twice a day; individuals between the ages of 50 and 69 years who consumed 900 mg of cocoa flavanols daily for three months improved brain performance on cognitive testing as well as on fMRI.[69] They had a "significant increase of" cerebral brain volume in the dentate gyrus "and the downstream subiculum in the body on the hippocampus." This indicates "that dietary cocoa flavanol consumption enhanced DG function in the aging human hippocampal circuit." Improvements "in the high-flavanol group was equivalent to improvements in cognition by approximately three decades of life."

In addition to taste, these are among the reasons that cocoa is one of my favorite nutrients. Three decades of cognition improvements is a significant neuroplastic purchase for consuming 450 mg of cocoa flavanols twice daily for three months. With artificial intelligence we could research timing of consuming the cocoa as well as dosage to maximize hacking neuroplasticity. Recording the time of day that the cocoa is ingested could be used to assess any potential impact on sleep.

There is growing evidence that cocoa and the flavonoids found in cocoa have significant positive effects on neurocognitive functions as well as improvements in mood. Thus, there is ample reason to appreciate cocoa flavonoids for their neuroprotective effects and to make them part of any protocol for improving brain health and brain power.

People with an average age of 72.9 were measured in their response to cocoa consumption for 24 hours and for 30 days.[70] This clinical trial was double blinded with one group receiving cocoa with 609 mg versus cocoa with 13 mg flavanols per serving. Trail Making Tests and Mini-Mental Status Exam were the tests of cognition. "Neurovascular coupling was associated with Trails B scores (p = 0.002) and performance on the 2-Back Task. Higher neurovascular coupling was also associated with significantly higher fractional anisotropy in cerebral white matter hyperintensities (p = 0.02)." After consuming cocoa for 30 days, both time for Trails B and neurovascular coupling improved in participants whose neurovascular coupling was impaired at baseline. Their improved neurovascular coupling was "also associated with greater white matter structural integrity."

A recent review[71] of the neuroprotective effects of the flavonoids in cocoa suggested that they "provoke angiogenesis, neurogenesis and changes in neuron morphology, mainly in regions involved in learning and memory." Another review[72] similarly found that cocoa flavonoids are neuroprotective and can enhance mood and cognitive function.

A review of studies suggested that another purchase of the flavonoids in cocoa might be their potential influence in protecting the supply of BDNF that nourishes individual brain cells, but more studies are needed to clarify this.[73] That might mean that cocoa could play multiple roles in protecting brain cells as well as bringing essential oxygen and nutrients to your hungry brain which only occupies about 2% of your total body mass while it uses about 20% of your body's energy. Since the flavonoids in cocoa are noted for powerful anti-inflammatory *and* antioxidant effects, this should make them telomere protectors.

These are added critical benefits of these flavonoids for the improved blood flow to the brain as well as increased health and flexibility of blood vessels. The flavanols protect your central nervous system from oxidative stress which causes injury; the heavy concentration of polyunsaturated fatty acids in the brain results in it being highly vulnerable to injury from reactive oxygen species or oxidative stress-mediated injuries. Cocoa flavonoids also suppress neuroinflammation which is associated with loss of brain integrity and performance.

Since many of us know the nuances of the several costs of sleep deprivation, it might be encouraging to know some of the associated physiology. One of those is reduced dilation of the blood vessels bringing nutrients to your brain. The profound cognitive functioning losses, especially in executive functions such as judgment, was observed in young healthy people who were deprived of sleep for one night. The good news in that study was that a serving of high-flavanol cocoa had restored good cardiovascular function, including better blood flow-mediated vasodilation, after two hours.

So, is cocoa in the early AM the best remedy for sleep deprivation, cognitive sluggishness and cardiovascular negative effects after a sleepless night? The jury is still out.

Stevia

But, if you choose a cup of cocoa for your wakeup call and want it to be sweet, remember to only sweeten your cocoa with stevia (no other sweetener) which is another of my favorite sources of polyphenols. That could have been influenced by the huge use of sugar in almost everything on the table of the Pennsylvania farm of my youth. Knowledge alone is not likely to modify most behaviors; that is to say, learning the science of the multiple negative impacts of consuming sugar was not, on its own, sufficient to change my food choices.

From a Positive Psychology perspective, it is likely that it was the tsunami of research on the health benefits of stevia that had the most impact on my move to shed sugar. Recently, for the first time in my life, I've walked into an ice cream store three separate times across several months *and* left those stores every time buying nothing. That was easy because the dessert recipes I make have several advantages; they (1) taste better; (2) will not interfere with my sleep; (3) energize me rather than leaving me feeling energy and cognitively depleted; (4) improve my cognitive function; AND (5) have the added mood elevating recognition that every molecule in my rendition of desserts and other foods is a molecule of physical, emotional, brain, and neurocognitive health.

These molecules include the health benefits of the polyphenols in stevia. A recent review of research on "steviol glycosides as natural sweeteners and molecules with therapeutic potential" found no harmful effects to human health.[74] Since it has no calories, does not cause dental caries, and is nonfermentive, it is a sweetener that is superior to sucrose and "scientific evidence encourages stevioside and rebaudioside A as sweetener alternatives to sucrose." Neither of these molecules are absorbed in the gastrointestinal tract. Gut microbiota action converts them to steviol. This research literature review also proposed that "these active compounds isolated from stevia rebaudiana possess interesting medicinal activities, including antidiabetic, antihypertensive, anti-inflammatory, antioxidant, anticancer, and antidiarrheal activity."

Some people have said they refuse to use stevia because they dislike its flavor. A physician colleague of mine had to grow her own stevia plants to be able to use it before it was "legal." There could be good news ahead on both of those issues. The European Food Safety Authority[75,76] twice wrote a "scientific opinion" on the "safety of 60 identified" components of stevia for use in food. They wrote that "the acceptable daily intake (ADI) of 4 mg/kg body weight (bw) per day will apply to all those steviol glycosides." Participants in a recent study rated Reb D and Reb A in ice creams "as being more sweet, pleasant, creamy, and milky, while Reb A was more artificial and chemical."[77]

With that many alternatives, hopefully food industries will increase use of a form of stevia, a natural sweetener, which more people can enjoy for the health of it. In addition to the regulation of sugar as suggested by Lustig,[3] a Positive Psychology approach could influence the food industry to increase the trend of using stevia for several reasons. First, since it is about 250–300 times sweeter than sucrose, cost of food production could be reduced by (1) needing less of the ingredient and (2) paying a lower price per product formulation. Second, they could improve their marketing by noting (1) reduced environmental impact AND (2) the associated health benefits of stevia. Despite the tsunami of research, adverse effects of stevia in humans have not been identified.

Stevia has had increasing interest in the scientific world. Although this natural sweetener has no calories while being 250–300 times sweeter than sucrose, it does not cause a spike in blood glucose which is followed by greater hunger. It has been shown to reduce calorie intake in two ways: (1) used in lieu of sugar, calories consumed are reduced and (2) when consumed in a beverage up to an hour prior to a meal, research participants significantly reduced their calories consumed in that meal even though study participants had free access to a variety of foods and were told to eat "until they felt comfortably full."[78] It also might help stabilize your sterile inflammation (inflammation not associated with an infection), blood glucose level, blood pressure, and the multiple consequences of obesity. It can increase beneficial and reduce harmful gut bacteria and could be beneficial for dental health.

Curcumin

Curcumin is the yellow pigment in turmeric, a spice used in diets of many cultures. I first came to value it in recipes from India that we enjoyed while living in Borneo. Indian curry has been a major part of my diet thereafter.

This pleiotropic polyphenol is neuroprotective in several ways.[79] It can influence cellular epigenetic changes and its anti-inflammatory effects include reducing neuroinflammation.[80] A recent review of studies suggested that curcumin may have "therapeutic potential in inflammatory diseases such as neurodegenerative diseases."[81]

Among its many potential benefits to your brain are improvements to your immune response as well as having anti-inflammatory, anticarcinogenic, and antioxidant effects. One of the neuroprotective influences found in rats was increased differentiation of neural stem cells into neurons.[82] It has shown a capacity to enhance neurogenesis and increase the number of neural stem cells in the hippocampus of adult mice.[83]

Working memory and sustained attention tasks were significantly improved in healthy humans aged 60–85 one hour after consuming 400 mg of curcumin.[84] After four weeks of this treatment, significant improvements were evident in mood and working memory.

Curcumin can improve the immune response.[85] It can have anti-inflammatory and antioxidant effects. It is believed to be neuroprotective and to promote remyelination. Absorption of curcumin could be increased by ingesting it with oil. In a personal communication with me a researcher suggested consuming curcumin with an oil might increase the absorption significantly. That makes taking it with a spoonful of olive oil, or guacamole, or avocado oil for the oleic acid, even more tasty. Again, this is only an example, not a suggestion.

Cinnamon

Cinnamon contains flavonoids, has anti-inflammatory influence, and is thought to be neuroprotective. In a systematic review of 40 studies,[86] most studies "proved that cinnamon significantly improves cognitive function (memory and learning)." Cinnamic acid, cinnamaldehyde, and eugenol are components of cinnamon which could be neuroprotective. They have anti-inflammatory, neurotrophic, and antioxidant effects. Eugenol and cinnamaldehyde can induce BDNF expression. Adding cinnamaldehyde or cinnamon to cells in vitro could "reduce tau. In vitro studies also showed that adding cinnamon or cinnamaldehyde to a cell medium can reduce tau aggregation, amyloid β and increase cell viability." A recent in vitro study identified cinnamaldehyde as providing this type of protection of "human neuroblastoma SH-SY5Y cell line" and these researchers opined that cinnamon polyphenols possess "multitarget anti" Alzheimer's disease activity.[87] Clinical studies reported mixed results. Further studies of the potential for cinnamon to prevent and reduce impaired cognitive functioning in humans are needed.

Research is beginning to explore components of cinnamon for their potential for improving recovery from traumatic brain injury. Some areas of the world have used cinnamon to treat various diseases. Cinnamon also appears to help stabilize blood sugar levels and increase insulin sensitivity. A review of random controlled trials, the gold standard of research, found that 1,480 people taking cinnamon as a supplement significantly reduced their body mass index (BMI), body weight, and their waist-to-hip ratio.[88] Other studies found cinnamon supplementation associated with reduced body fat mass in adults after consuming 2 grams of cinnamon per day for 12 weeks. One reviewer suggested research to see what 3 grams of powdered cinnamon per day would influence benefits of these flavonoids.

Adding cinnamon to many foods can be a delicious way to promote healthy brain aging. Also, I make cinnamon tea by pouring boiling water on a large amount of cinnamon and letting it rest for a while before refrigeration.

Resveratrol

Resveratrol is another polyphenol that is neuroprotective. It is associated with living longer while preserving memory and hippocampal microstructure.[30,89] Chen and colleagues[90] reviewed studies that found anti-inflammatory, antioxidant, and immune regulating functions of resveratrol and opined that these benefits might be "attributed to resveratrol's role in protecting the intestinal barrier, regulating the gut microbiome, and inhibiting intestinal inflammation." This polyphenol occurs naturally in grapes, purple grape juice, and some berries such as blueberries and cranberries.

A recent review[91] of studies on the effects of blueberries and their polyphenols found that "blueberries may play a role in the improvement of markers of vascular function. Their effects were observed following both post-prandial and long-term consumption, particularly in subjects with risk factors and/or disease conditions." More human studies are needed to further clarify these vascular benefits and how this might enhance brain health.

Elders aged 50–80 years had better brain function on neuroimaging and better memory after 12 weeks of consuming "cranberry supplementation (equivalent to 1 small cup of cranberries)."[92] Neuroimaging found "increased regional perfusion in the right entorhinal cortex, the accumbens area and the caudate" of individuals consuming cranberry supplement.

Unfortunately, resveratrol is poorly absorbed by humans. Fortunately, once it gets absorbed into your bloodstream, it can cross the blood-brain barrier where it can improve vascular function and activate longevity genes[93] and minimize brain aging.

Quercetin

Quercetin occurs naturally in many fruits and plants. It protects against neuroinflammation and oxidative stress. A recent review of research findings suggests that quercetin may have "therapeutic potential against psychological and neurodegenerative disorders."[94] Neuron survival may be enhanced by its influence on neurotrophic factors.

It has been suggested that increased foods rich in flavonoids such as quercetin could improve arterial health and function across the lifespan. Polyphenol and flavonoid rich foods in addition to cocoa and stevia discussed above include berries,[91] pomegranate,[95] moringa,[96] ginger, tea, and some spices.[93] Moringa is native to India; an important flavonoid in its leaves is quercetin.

CoQ$_{10}$

Another important antioxidant is CoQ$_{10}$ which could improve mitochondrial function which is particularly important with Alzheimer's disease and other neurodegenerative diseases.[97] Our cells produce some of this. Good food sources include fatty fish, legumes, and nuts. Tofu and sardines are my favorite sources.

HLXL

Another interesting study involved "a Chinese medicine (HLXL), which contains 56 bioactive natural products identified in 11 medicinal plants and displays potent anti-inflammatory and immuno-modulatory activity."[98] Both in vitro and in vivo studies with HLXL found "significantly reduced amyloid neuropathology." These researchers concluded that it "is conceivable that the traditional wisdom of natural medicine in combination with modern science and technology would be the best strategy in developing effective therapeutics for AD" with AD meaning Alzheimer's disease. This is another excellent contribution to our database by Rudolph Tanzi in whose lab the first Alzheimer's gene was identified.

Discussion

Artificial intelligence used with human creativity, judgment, reasoning, and guidance could expedite and refine the focus of rapidly evolving neuroscience of dietary impacts on our Intelligent Neuroplasticity Revolution to enhance human brain power and healthy aging. It is timely and urgent to establish personalized goals and strategies to maximize the best brain chemistry, architecture, and function at any age for all our global citizens with a **Diet to Reduce Neuroinflammation and Nourish Your Building Brain**.

This is an extensive overview of evolving neuroscience on nutrients to build, enhance, and protect brain health. This delicious topic has enough evidence-based interventions to fill an entire very large book. Herein the basics addressed include an emphasis that it is imperative to maximize nutritional purchases for brain health from preconception through end of life. It considers how artificial intelligence can help personalize nutrition considering the many unique aspects of current physical health, cultural factors, resources, and the complexities of nutrient interactions, the gut microbiota, the gut-brain axis, and enjoying eating for the health of it. This necessarily includes a balance of energy consumed and energy used in ways that influence autophagy for increasing the health of cells.

As described above, autophagy may become our favorite four-syllable word. Not only is it an elite recycling system on the molecular level. It refurbishes cells by degrading and removing damaged and dysfunctional components of cells. There is some research suggesting that increasing autophagy in the hippocampus might help improve memory. Although this research was in vitro and with other animals, it is an important field of study that could increase options for improving memory.

Autophagy is influenced by the balance of energy consumed and the energy being used so it can be increased as easily as combining time-restricted eating with a regular exercise routine. That's why my exercise routine precedes any calory intake in the early morning.

Nutrition, diets, and hydration from in utero and across the entire lifespan play a central part in the development of several age-related diseases, such as cardiovascular disease, neurodegenerative disease, and dementia. Fat cells can increase silent inflammation. Neuroinflammation is a major risk factor for dementia. Asymptomatic inflammation is a health hazard that can be addressed by oral intake in addition to other evidence-based interventions such as exercise. Well-managed nutritional intake adapted to age is essential to avoid silent inflammation, maintain mental freshness, and provide a good quality of life at any age.

Diets also influence the process of aging itself. By adhering to certain diets, such as the Okinawan and Mediterranean diets, the onset of age-related diseases can be prevented or delayed. This is achieved directly by certain elements included in the diet, such as fruits, vegetables, and omega-3 fatty acids, and indirectly by their positive effect on bodyweight management. Moreover, the introduction of fasting episodes into our diet may also influence autophagy as well as contributing to healthier aging. Diet recommendations change with age, and this must be taken into consideration when developing a diet tailored to the needs of all people of all ages. Future clinical and preclinical investigations are expected to lead to novel nutritional guidelines for the entire lifespan in the future.

Calorie restriction with adequate nutrients has been associated with health benefits through increased longevity in organisms from yeast to flies, worms, and mammals. Research suggests that *hara hachi bu*, or "eat until you are 80% full," has been an important factor in exceptional longevity with increased health span for one human population to the extent that their country was considered a centenarian center of the world.[8] Reducing calories by 30% was associated with an average of 20% improvement in verbal memory after three months.[99] Some of these cognitive and general health benefits of calorie restriction in humans are thought to be related to the reduction of inflammation and oxidative damage. The brain and body interaction in response to the restriction of calories could influence the sleep cycle which has been associated with inflammation; and it might be related to "the global metabolic reprogramming of the central biological clock."[20] Still, the molecular nuances of calorie restriction remain poorly understood. The problem with low-grade inflammation that can be chronic as well as undiagnosed is that it can be associated with decreased cognitive functioning with aging.[100,101]

A research review[102] supports the value of Ramadan fasting and the benefit of time-restricted eating to improve composition of gut microbiota. However, the essential factor influencing a healthy gut microbiota is diet.

Intermittent fasting in animals found benefits that equaled or exceeded the benefits of calorie restriction; brain cells in these animals were more capable of resisting the injury of an injection into the hippocampus that has known toxic effects.[103] Reducing caloric intake seems to improve synaptic resilience to damage and modify the number, architecture and performance of synapses.[104] A reduction in inflammation with better preservation of cognitive function in animals with sepsis suggested that intermittent fasting can induce adaptive responses systemically as well as in the brain.[105]

Neither calorie restriction nor intermittent fasting should be taken lightly. Each requires healthy nutrition. Whether that uses the Okinawan diet, which was one of the factors in the lives of high-functioning centenarians,[8] or the Mediterranean diet, which has some evidence for being "a potential strategy to reduce cognitive decline in older age,"[106] is discretionary. Since both emphasize vegetables, fruits, fish as a source of protein, and low glycemic load, both would be rich in polyphenols and healthier polyunsaturated fats and would have antioxidant and anti-inflammatory benefits.

A research review suggests that a high brain concentration of DHA, an essential omega-3 fatty acid, can optimize synaptic plasticity and efficiency and help maintain homeostasis in the synapses.[107] For efficient transmission of data between brain cells, the plasma membrane must remain fluid. DHA is a component of this membrane. Adequate intake of the essential fatty acids is crucial to maintaining the fluid transmission of molecules across neuronal membranes because this is where much of the action takes place for such core brain functions as learning, memory and sleep.[108] They are also essential in building the myelin sheath that enhances efficient processing of information.[109]

That's why I offer to share my sardines with friends. DHA for me and for thee.

The finding that DHA is vulnerable to oxidative damage underscores both the need for polyphenols as well as the complexity of the interactive neuroplastic influence of the several components of the dietary intervention matrix which also needs to consider essential vitamins and minerals. For example, since brain health requires adequate Vitamin B_{12}, episodic measures of this status are recommended.[110]

Both the Okinawan and the Mediterranean diets are associated with better brain health as both diets include plenty of fruits and vegetables; essential omega-3 fatty acids from fish as a source of protein; and fiber in the plant foods which aids in digestion of the many polyphenols while also stabilizing the gut microbiota. Having Okinawa considered a centenarian center of the world until Western dietary influences changed oral intake can strengthen our resolve to learn from the wisdom of their aged and learn from their loss of status as a centenarian center of the world when they took on Western ways of consuming. Let that motivate us to personalize nutrition for healthy aging with foods seen in Figure 7.2 to Reduce Neuroinflammation and Nourish Your Building Brain.

We can eat and fast together for hacking neuroplasticity. We can use artificial intelligence and the Intelligent Neuroplasticity Revolution to Help Your Healthy Aging Brain.

That begins with and emphasizes the basic tenet that energy consumed approximates energy used. As with all aspects of nutrition, that varies with individual status including age, physical health, etc. Dietary

Figure 7.2
This illustration by David Yu, MD, shows a cornucopia of dietary choices you have for reducing neuro-inflammation and nourishing your hungry brain.

Guidelines for Americans, 2020–2025, 9th Edition is available at https://www.dietaryguidelines.gov/sites/default/files/2020-12/Dietary_Guidelines_for_Americans_2020-2025.pdf and can influence factors used by artificial intelligence to personalize oral intake to maximize best brain chemistry, architecture, and performance across the entire lifespan.

A review of digital technology with cognitively impaired patients and their care providers emphasized the "need for a theoretical framework to conceptualize and govern" behavioral interventions using technology.[111] Researchers[112] opined that current wearables for 24-hour assessment of physical behaviors for "intensity, posture/activity type, and biological state" are highly variable in design and have poor quality methodology.

Artificial intelligence used with human creativity, problem solving, judgment, and oversight can increase frequency of making personalized healthy dietary choices in several ways. Creating wearables for providing unpredictable positive reinforcement of the individual's activities after they begin and after they finish activities would be likely to increase those behaviors. It would also be valuable to provide prompts some moments after an activity was scheduled if it has not occurred. Healthcare providers who participated in setting these personalized goals and strategies might also be provided with a summary on these specifics to help enhance adherence as well as enjoyment of a **Diet to Reduce Neuroinflammation and Nourish Your Building Brain.**

References

1. Watso JC & Farquhar WB. (2019). Hydration Status and Cardiovascular Function. *Nutrients*, 11: 1866.
2. Lauriola M, Mangiacotti A, D'Onofrio G, Cascavilla L, Paris F, Paroni G, Seripa D, Greco A & Sancarlo D. (2018). Neurocognitive Disorders and Dehydration in Older Patients: Clinical Experience Supports the Hydromolecular Hypothesis of Dementia. *Nutrients*, 10: 562.
3. Lustig RH. (2020). Ultraprocessed Food: Addictive, Toxic, and Ready for Regulation. *Nutrients*, 12: 3401. DOI: 10.3390/nu12113401
4. Exclusive Breastfeeding for Optimal Growth, Development and Health of Infants (who.int).
5. Kafouri S, Kramer M, Leonard G, Perron M, Pike B, Richer L, et al. (2013). Breastfeeding and Brain Structure in Adolescence. *Intern J Epidemiology*, 42: 150–159. DOI: 10.1093/ije/dys172
6. WHO Obtained 3/3/2024: Infant and Young Child Feeding (who.int).
7. Galiè S, Canudas S, Muralidharan J, García-Gavilán J, Bulló M, Salas-Salvadó J, et al. (2020). Impact of Nutrition on Telomere Health: Systematic Review of Observational Cohort Studies and Randomized Clinical Trials. *Adv Nutr*, 11: 576–601. DOI: 10.1093/advances/nmz107
8. Willcox DC, Scapagnini G & Willcox BJ. (2014). Healthy Aging Diets Other Than the Mediterranean: A Focus on the Okinawan Diet. *Mech Ageing Development*, 136–137: 148–162.
9. Jacka FN, Cherbuin N, Anstey KJ, Sachdev P & Butterworth P. (2015). Western Diet Is Associated with a Smaller Hippocampus: A Longitudinal Investigation. *BMC Med*, 13: 215, DOI: 10.1186/s12916-015-0461-x
10. Brunt VE, LaRocca TJ, Bazzoni AE, Sapinsley ZJ, Miyamoto-Ditmon J, Gioscia-Ryan RA, Neilson AP, et al. (2021). The Gut Microbiome-Derived Metabolite Trimethylamine N-Oxide Modulates Neuroinflammation and Cognitive Function with Aging. *Geroscience*, 43: 377–394.
11. Chaudhari DS, Jain S, Yata VK, Mishra SP, Kumar A, Fraser A, Kociolek J, et al. (2023). Unique Trans-Kingdom Microbiome Structural and Functional Signatures Predict Cognitive Decline in Older Adults. GeroScience. DOI: 10.1007/s11357-023-00799-1

12. Nagpal R, Neth BJ, Wang S, Craft S & Yadav H. (2019). Modified Mediterranean-Ketogenic Diet Modulates Gut Microbiome and Short-Chain Fatty Acids in Association with Alzheimer's Disease Markers in Subjects with Mild Cognitive Impairment. *EBioMed*, 47: 529–542.

13. Kasti AN, Nikolaki MD, Synodinou KD, Katsas KN, Petsis K, Lambrinou S, Pyrousis IA, Triantafyllou K, et al. (2022). The Effects of Stevia Consumption on Gut Bacteria: Friend or Foe? *Microorganisms*, 10: 744. doi: 10.3390/microorganisms10040744.

14. Sun Y, Baptista LC, Roberts LM, Jumbo-Lucioni P. McMahon LL, Buford TW, Carter CS, et al. (2020). The Gut Microbiome as a Therapeutic Target for Cognitive Impairment. *J Gerontol A Biol Sci Med Sci*, 75: 1242–1250. doi: 10.1093/gerona/glz281.

15. Chlebowski RT, Rapp S, Aragaki AK, Pan K & Neuhouser ML. (2020). Low-Fat Dietary Pattern and global cognitive function: Exploratory Analyses of the Women's Health Initiative (WHI) Randomized Dietary Modification Trial. *EClinical Med*, 18: 100240. DOI: 10.1016/j.eclinm.2019.100240

16. Agarwal P, Leurgans SE, Agrawal S, Aggarwal NT, Cherian LJ, James BD, Dhana K, Barnes LL, Bennett DA, Schneider JA, et al. (2023) Association of Mediterranean-DASH Intervention for Neurodegenerative Delay and Mediterranean Diets with Alzheimer Disease Pathology. *Neurology*, 100(22): e2259–e2268. doi: 10.1212/WNL.0000000000207176. Epub 2023 Mar 8.

17. Kandel P, Semerci F, Mishra R, Choi W, Baji A, Baluy D, Ma L, et al. (2022). Oleic Acid Is an Endogenous Ligand of TLX/NR2E1 That Triggers Hippocampal Neurogenesis. *PNAS*, 119(13): e2023784119. DOI: 10.1073/pnas.2023784119

18. Peterson C. (2008) What Is Positive Psychology, and What Is It Not? *Psychol Today*, May 16.

19. Li C, Luan Z, Zhao Y, Chen J, Yang Y, Wang C, Jing Y, et al. (2022). Deep Insights into the Gut Microbial Community of Extreme Longevity in South Chinese Centenarians by Ultra-Deep Metagenomics and Large-Scale Culturomics. *NPJ Biofilms Microbiomes*, 8: 28. doi: 10.1038/s41522-022-00282-3.

20. Fusco S & Pani P. (2013). Brain Response to Calorie Restriction. *Cell Mol Life Sci*, 70: 3157–3170. DOI: 10.1007/s00018-012-1223-y

21. Mattson MP, Moehl K, Ghena N, Schmaedick M & Cheng A. (2018). Intermittent Metabolic Switching, Neuroplasticity and Brain Health. *Nat Rev Neurosci*, 19: 63–80.

22. Glatigny M, Moriceau S, Rivagorda M, Ramos-Brossier M, Mascimbeni AC, Lante F, et al. (2019). Autophagy Is Required for Memory Formation and Reverses Age-Related Memory Decline. *Curr Biol*, 29: 435–448. DOI: 10.1016/j.cub.2018.12.021

23. Anderson RM & Weindruch R. (2012) The Caloric Restriction Paradigm: Implications for Healthy Human Aging. *Amer J Human Biol*, 24: 101–106.

24. Ezzati A & Pak VM. (2023). The Effects of Time-Restricted Eating on Sleep, Cognitive Decline, and Alzheimer's Disease. *Exper Geron*, 171: 112033. doi: 10.1016/j.exger.2022.112033. Epub 2022 Nov 17.

25. Longo VD & Anderson RM. (2022). Nutrition, Longevity and Disease: From Molecular Mechanisms to Interventions. *Cell*, 185: 1455–1470.
26. de Cabo R & Mattson MP. (2019). Effects of Intermittent Fasting on Health, Aging, and Disease. *N Engl J Med*, 381: 2541–2551. DOI: 10.1056/NEJMra1905136
27. Balasubramanian P, DelFavero J, Ungvari A, Papp M, Tarantini A, Price N, de Cabo R & Tarantini S. (2020). Time-Restricted Feeding (TRF) for Prevention of Age-Related Vascular Cognitive Impairment and Dementia. *Ageing Res Rev*, 64: 101189. DOI: 10.1016/j.arr.2020.101189
28. Rahbar AR, Safavi E, Rooholamini M, Jaafari F, Darvishi S & Rahbar A. (2019). Effects of Intermittent Fasting during Ramadan on Insulin-Like Growth Factor-1, Interleukin 2, and Lipid Profile in Healthy Muslims. *Int J Prev Med*, 10: 7.
29. Gomez-Pinilla F & Tyagi, E. (2013). Diet and Cognition: Interplay between Cell Metabolism and Neuronal Plasticity. *Curr Opin Clin Nutr Metab Care*, 16(6): 726–733.
30. Dauncey MJ. (2014). Nutrition, the Brain and Cognitive Decline: Insights from Epigenetics. *Eur J Clin Nutr*, 68: 1179–1185. DOI: 10.1038/ejcn.2014.173
31. Murphy T, Dias GP & Thuret S. (2014). Effects of Diet on Brain Plasticity in Animal and Human Studies: Mind the Gap. Neural Plast, 563160. doi: 10.1155/2014/563160. Epub 2014 May 1
32. Chesnokova V, Pechnick RN & Wawrowsky K. (2016). Chronic Peripheral Inflammation, Hippocampal Neurogenesis, and Behavior. *Brain, Behavior Immunity*. DOI: 10.1016/j.bbi.2016.01.017
33. Barbaresko J, Koch M, Schulze MB & Nöthlings U. (2013). Dietary Pattern Analysis and Biomarkers of Low-Grade Inflammation: A Systematic Literature Review. *Nutr Rev*, 71(8): 511–527.
34. Georgieff MK. (2023). Early Life Nutrition and Brain Development: Breakthroughs, Challenges, and New Horizons. *Proc Nutr Soc*, 82(2): 104–112. DOI: 10.1017/S0029665122002774
35. Fekete M, Szarvas Z, Fazekas-Pongor V, Feher A, Csipo T, Forrai J, Dosa N, et al. (2023). Nutrition Strategies Promoting Healthy Aging: From Improvement of Cardiovascular and Brain Health to Prevention of Age-Associated Diseases. *Nutrients*, 15: 47. DOI: 10.3390/nu15010047
36. Raghavan G, Bapna A, Mehta A, Shah A & Vyas T. (2023). Effect of Sugar Replacement with Stevia-Based Tabletop Sweetener on Weight and Cardiometabolic Health among Indian Adults. *Nutrients*, 15: 1744. DOI: 10.3390/nu15071744
37. Tur JA & Bibiloni MDM. (2019) Anthropometry, Body Composition and Resting Energy Expenditure in Human. *Nutrients*, 11: 1891.
38. Wang H, Lv Y, Ti G & Ren G. (2022). Association of Low-Carbohydrate-Diet Score and Cognitive Performance in Older Adults: National Health and Nutrition Examination Survey (NHANES). *BMC Geriatr*, 22: 983. doi: 10.1186/s12877-022-03607-1.
39. WHO Healthy Diet Fact Sheet. https://www.who.int/news-room/fact-sheets/detail/healthy-diet

40. Barberger-Gateau P. (2014). Nutrition and Brain Aging: How Can We Move Ahead? *Eur J Clin Nutr*, 68: 1245–1249. https://doi.org/10.1038/ejcn.2014.177

41. Raji C, Erickson KI, Lopez OL, Kuller LH, Gach HM, Thompson PM, et al. (2014). Regular Fish Consumption and Age-Related Brain Gray Matter Loss. *Am J Prev Med*, 47(4): 444–451. DOI: 10.1016/j.amepre.2014.05.037

42. Schaefer EJ., Bongard V, Beiser AS, Lamon-Fava S, Robins SJ, Au R, et al. (2006). Plasma Phosphatidylcholine Docosahexaenoic Acid Content and Risk of Dementia and Alzheimer Disease: The Framingham Heart Study. *Arch Neurol*, 63: 1545–1550.

43. Sambra V, Echeverria F, Valenzuela A, Chouinard-Watking R & Valenquela R. (2021). Docosahexaenoic and Arachidonic Acids as Neuroprotective Nutrients throughout the Life Cycle. Docosahexaenoic and Arachidonic Acids as Neuroprotective Nutrients throughout the Life Cycle. *Nutrients*, 13: 986. DOI: 10.3390/nu13030986

44. Boespflug EL, McNamara RK, Eliassen, JC, Schidler MD & Krikorian R. (2016). Fish Oil Supplementation Increases Event-Related Posterior Cingulate Activation in Older Adults with Subjective Memory Impairment. *J Nutr Health & Aging*, 20(2): 161–169.

45. Satizabal CL, Himali JJ, Beiser AS, Ramachandran V, van Lent DM, Himali D, Aparicio HJ, et al. (2022). Association of Red Blood Cell Omega-3 Fatty Acids with MRI Markers and Cognitive Function in Midlife: The Framingham Heart Study. *Neurology*, 99(23): e2572–e2582. DOI: 10.1212/WNL.0000000000201296

46. Sun Y, Liu B, Snetselaar LG, Wallace RB, Shadyab AH, Kroenke CH, Haring B, et al. (2021). Association of Major Dietary Protein Sources with All-Cause and Cause-Specific Mortality: Prospective Cohort Study. *J Am Heart Assoc*, 10: e015553. DOI: 10.1161/JAHA.119.015553

47. Ali MA, Poortvliet E, Stromberg R & Yngve A. (2011). Polyamines in Foods: Development of a Food Database. *Food & Nutr Res*, 55: 5572. doi: 10.3402/fnr.v55i0.5572.

48. Amri Z, Amor IB, Zarrouk A, Chaaba R, Gargouri J, Hammami M, Hammami S, et al. (2022). Anti-Glycation, Antiplatelet and Antioxidant Efects of Diferent Pomegranate Parts. *BMC Compl Med & Ther*, 22: 339. DOI: 10.1186/s12906-022-03824-6

49. Barber TM, Kabisch S, Pfeiffer AFH & Weickert MO. (2020). The Health Benefits of Dietary Fibre. *Nutrients*, 12: 3209. DOI: 10.3390/nu12103209

50. Glabska D, Guzek D, Groele B & Gutkowska K. (2020). Fruit and Vegetable Intake and Mental Health in Adults: A Systematic Review. *Nutrients*, 12: 115.

51. Zhang H, Tian W, Qi G & Sun Y. (2022). Hypertension, Dietary Fiber Intake, and Cognitive Function in Older Adults [from the National Health and Nutrition Examination Survey Data (2011–2014)]. *Front Nutr*, 9: 1024627. doi: 10.3389/fnut.2022.1024627. eCollection 2022.

52. Baker LD, Manson JE, Rapp SR, Sesso HD, Gaussoin SA, Shumaker SA, Espeland MA, et al. (2022). Effects of Cocoa Extract and a Multivitamin on Cognitive Function: A Randomized Clinical Trial. *Alzheimers Dement*, 19(4): 1308–1319. DOI: 10.1002/alz.12767

53. Yeung L-K, Alschuler DM, Wall M, Luttman-Gibson H, Copeland T, Hale C, Sloan RP, et al. (2023). Multivitamin Supplementation Improves Memory in Older Adults: A Randomized Clinical Trial. Amer J Clin Nutr. DOI: 10.1016/j.ajcnut.2023.05.011

54. Lauer AA, Grimm HS, Apel B, Golobrodska N, Kruse L, Ratanski E, Schulten N, et al. (2022). Mechanistic Link between Vitamin B12 and Alzheimer's Disease. Biomol, 12: 129. DOI: 10.3390/biom12010129

55. Baltrusch, S. (2021). The Role of Neurotropic B Vitamins in Nerve Regeneration. Biomed Res Int, 9968228. DOI: 10.1155/2021/9968228

56. Oulhaj A, Jerneren F, Refsum H, Smith AD & de Jager CA. (2016). Omega-3 Fatty Acid Status Enhances the Prevention of Cognitive Decline by b Vitamins in Mild Cognitive Impairment. J Alzheimers Dis, 50: 547–557.

57. Jernerén F, Elshorbagy AK, Oulhaj A, Smith SM, Refsum H, Smith AD, et al. (2015) Brain Atrophy in Cognitively Impaired Elderly: The Importance of Long-Chain Omega-3 Fatty Acids and B Vitamin Status in a Randomized Controlled Trial. Am J Clin Nutr, 102: 215–221.

58. Chimakurthy AK, Lingam S, Pasya SKR & Copeland BJ. (2023). A Systematic Review of Dietary Supplements in Alzheimer's Disease. Cureus, 15(1): e33982. doi: 10.7759/cureus.33982. eCollection 2023 Jan.

59. Tramontin Nd S, Luciano TF, Marques Sc O, de Souza CT & Muller AP. (2020). Ginger and Avocado as Nutraceuticals for Obesity and Its Comorbidities. Phytother Res, 34: 1282–1290.

60. Liu Y, Deng SJ, Zhang Z, Gu Y, Xia SN, Bao XY, Cao X, Xu Y, et al. (2020). 6-Gingerol Attenuates Microglia-Mediated Neuroinflammation and Ischemic Brain Injuries through Akt-mTOR-STAT3 Signaling Pathway. E J Phar, 883: 173294. DOI: 10.1016/j.ejphar.2020.173294

61. Cheng FW, Ford NA & Taylor MK. (2021) US Older Adults That Consume Avocado or Guacamole Have Better Cognition Than Non-consumers: National Health and Nutrition Examination Survey 2011–2014. Front Nutr, 8: 746453. DOI: 10.3389/fnut.2021.746453

62. Zhai Y, Liu B-G, Mo X-N, Zou M, Mei X-P, Chen W, Huang G-D, Wu L, et al. (2021). Gingerol Ameliorates Neuronal Damage Induced by Hypoxia-Reoxygenation Via the miR-210/Brain-Derived Neurotrophic Factor Axis. Kaohsiung J Med Sci, 1–11. doi: 10.1002/kjm2.12486. Epub 2021 Dec 28.

63. Sarubbo F, Moranta D. Tejada S, Jiménez M & Esteban S. (2023). Impact of Gut Microbiota in Brain Ageing: Polyphenols as Beneficial Modulators. Antioxidants, 12: 812. DOI: 10.3390/antiox12040812

64. Sarubbo F, Cavallucci V & Pani G. (2022). The Influence of Gut Microbiota on Neurogenesis: Evidence and Hopes. Cells, 11, 382.

65. Holland TM, Agarwal P, Wang Y, Dhana K, Leurgans SE, Shea K, Booth SL, et al. (2022). Association of Dietary Intake of Flavonols with Changes in Global Cognition and Several Cognitive Abilities. NeurPublish. DOI: 10.1212/WNL.0000000000201541

66. Monahan KD. (2012). Effect of Cocoa/Chocolate Ingestion on Brachial Artery Flow-Mediated Dilation and Its Relevance to Cardiovascular Health and Disease in Humans. Arch Biochem Biophys, DOI: 10.1016/j.abb.2012.02.021

67. Monahan KD, Feehan RP, Kunselman AR, Preston AG, Miller DL, Lott MEJ, et al. (2012). Dose-Dependent Increases in Flow-Mediated Dilation Following Acute Cocoa Ingestion in Healthy Older Adults. *J Applied Physiol*, 111: 1568–1574. DOI: 10.1152/japplphysiol.00865.2011
68. Nishiwaki M, Nakano Y & Matsumoto N. (2019). Effects of Regular High-Cocoa Chocolate Intake on Arterial Stiffness and Metabolic Characteristics during Exercise. *Nutrition*, 60: 53–58. DOI: 10.1016/j.nut.2018.09.021
69. Brickman AM, Khan UA, Provenzano FA, Yeung L-K, Suzuki W, Schroeter H, et al. (2014). Enhancing Dentate Gyrus Function with Dietary Flavanols Improves Cognition in Older Adults. Nat Neurosci. DOI: 10.1038/nn.3850
70. Sorond FA, Hurwitz S, Salat DH, Greve DN & Fisher ND. (2013). Neurovascular Coupling, Cerebral White Matter Integrity, and Response to Cocoa in Older People. *Neurology*, 81: 904–909.
71. Nehlig A. (2013). The Neuroprotective Effects of Cocoa Flavanol and Its Influence on Cognitive Performance. *Br J Clin Pharmacol*, 75(3): 716–727.
72. Latif, R. (2013). Health Benefits of Cocoa. *Curr Opin Clin Nutr Metab Care*, 16: 669–674.
73. Martin MA & Ramos S. (2021). Impact of Cocoa Flavanols on Human Health. *Food and Chem Toxicol*, 151: 112121. DOI: 10.1016/j.fct.2021.112121
74. Orellana-Paucar AM. (2023). Steviol Glycosides from Stevia Rebaudiana: An Updated Overview of Their Sweetening Activity, Pharmacological Properties, and Safety Aspects. *Molecules*, 28: 1258. doi: 10.3390/molecules 28031258.
75. EFSA FAF Panel (EFSA Panel on Food Additives and Flavourings), Younes M, Aquilina G, Engel K-H, Fowler PJ, Frutos Fernandez MJ, Furst P, G € urtler R, Gundert-Remy U, Husøy T, Manco M, Mennes W, Moldeus P, Passamonti S, Shah R, Waalkens-Berendsen I, Wright M, Barat Baviera JM, Degen G, Herman L, Leblanc J-C, Wolfle D, Aguilera J, Giarola A, Smeraldi C, Vianello G, Castle L, et al. (2022). Scientific Opinion on the Safety of the Proposed Amendment of the Specifications for Enzymatically Produced Steviol Glycosides (E 960c): Rebaudioside D Produced Via Enzymatic Bioconversion of Purified Stevia Leaf Extract. *EFSA J*, 20(5): 7291. 23 pp. doi: 10.2903/j.efsa.2022.7291
76. EFSA FAF Panel (EFSA Panel on Food Additives and Flavourings), Younes M, Aquilina G, Engel K-H, Fowler P, Frutos Fernandez MJ, Furst P, Gurtler R, Gundert-Remy U, Husøy T, Manco M, Mennes W, Moldeus P, Passamonti S, Shah R, Waalkens-Berendsen I, Wolfle D, Wright M, Degen G, Giarola A, Rincon AM, Castle L, et al. (2020). Scientific Opinion on the Safety of a Proposed Amendment of the Specifications for Steviol Glycosides (E 960) as a Food Additive: To Expand the List of Steviol Glycosides to All Those Identified in the Leaves of Stevia Rebaudiana Bertoni. *EFSA J*, 18(4): 6106, 32 pp. doi: 10.2903/j.efsa.2020.6106
77. Muenprasitivej N, Tao R, Nardone SJ & Cho S. (2022). The Effect of Steviol Glycosides on Sensory Properties and Acceptability of Ice Cream. *Foods*, 11: 1745. DOI: 10.3390/foods11121745

78. Stamataki NS, Scott C, Elliott R, McKie S, Bosscher D, McLaughlin JT, et al. (2020). Stevia Beverage Consumption prior to Lunch Reduces Appetite and Total Energy Intake without Affecting Glycemia or Attentional Bias to Food Cues: A Double-Blind Randomized Controlled Trial in Healthy Adults. J Nutr. DOI: 10.1093/jn/nxaa038

79. Pogacnik L, Ota A & Ulrih NP. (2020). An Overview of Crucial Dietary Substances and Their Modes of Action for Prevention of Neurodegenerative Diseases. *Cells*, 9: 576; DOI: 10.3390/cells9030576

80. Benameur T, Giacomucci G, Panaro MA, Ruggiero M, Trotta T, Monda V, Pizzolorusso I, et al. (2022). New Promising Therapeutic Avenues of Curcumin in Brain Diseases. *Molecules*, 27: 236. DOI: 10.3390/molecules27 010236

81. Benameur T, Frota Gaban SV, Giacomucci G, Filannino FM, Trotta T, Polito R, Messina G, et al. (2023). The Effects of Curcumin on Inflammasome: Latest Update. *Molecules*, 28: 742. doi: 10.3390/molecules28020742.

82. Chen F, Wang H, Xiang X, Yuan J, Chu W, Xue X, Zhu H, et al. (2014). Curcumin Increased the Differentiation Rate of Neurons in Neural Stem Cells Via Wnt Signaling in Vitro Study. *J Surgical Res* xxx, 1–7. DOI: 10.1016/j. jss.2014.06.026

83. Kim JJ, Son TG, Park HR, Park M, Kim M, Kim HS, et al. (2008). Curcumin Stimulates Proliferation of Embryonic Neural Progenitor Cells and Neurogenesis in the Adult Hippocampus. *J Biol Chem*, 2839(21): 14497–14505.

84. Cox, Katherine HM, Pipingas A & Scholey AB. (2015). Investigation of the Effects of Solid Lipid Curcumin on Cognition and Mood in a Healthy Older Population. *J Psychopharmacol*, 29(5): 642–651.

85. Ganjali S, Sahebkar A, Mahdipour E, Jamialahmadi K, Toraki S, Akhlaghi S, Ferns G, et al. (2014). Investigation of the Effects of Curcumin on Serum Cytokines in Obese Individuals: A Randomized Controlled Trial. *Sci World J*: Article ID 898361. DOI: 10.1155/2014/898361

86. Nakhaee S, Kooshki A, Hormozi A, Akbari A, Mehrpour O & Farrokhfall K. (2024). Cinnamon and Cognitive Function: A Systematic Review of Preclinical and Clinical Studies. *Nutr Neurosci*. DOI: 10.1080/1028415X.2023. 2166436

87. Ciaramelli C, Palmioli A, Angotti I, Colombo L, De Luigi A, Sala G, Salmona M & Airoldi C (2022). NMR-Driven Identification of Cinnamon Bud and Bark Components with AntiAβ Activity. *Front Chem*, 10: 896253. DOI: 10.3389/fchem.2022.896253

88. Yazdanpanah A-YM, Hooshmandi H, Ramezani-Jolfaie N & Salehi-Abargouei A. (2020). Effects of Cinnamon Supplementation on Body Weight and Composition in Adults: A Systematic Review and meta-Analysis of Controlled Clinical Trials. *Phytotherapy Res*, 34: 448–463. doi: 10.1002/ptr.6539

89. Witte AV, Kerti L, Margulies CS & Floel A. (2014). Effects of Resveratrol on Memory Performance, Hippocampal Functional Connectivity, and Glucose Metabolism in Healthy Older Adults. *J Neurosci*, 34(23): 7862–7870.

90. Chen X, Zhang J, Yin N, Wele P, Li F, Dave S, et al. (2023). Resveratrol in Disease Prevention and Health Promotion: A Role of the Gut Microbiome. *Crit Rev Food Sci Nutr*, Jan 2; 1–18. doi: 10.1080/10408398.2022. 2159921

91. Martini D, Marino M, Venturi S, Tucci M, Zacas DK, Riso P, Purrini M, Bo CD, et al. (2023). Blueberries and Their Bioactives in the Modulation of Oxidative Stress, Inflammation and Cardio/Vascular Function Markers: A Systematic Review of Human Intervention Studies. J Nutr Biochem, 111: 109154. DOI: 10.1016/j.jnutbio.2022.109154

92. Flanagan E, Cameron D, Sobhan R, Wong C, Pontifex MG, Tosi N, Mena P, et al. (2022). Chronic Consumption of Cranberries (Vaccinium macrocarpon) for 12 Weeks Improves Episodic Memory and Regional Brain Perfusion in Healthy Older Adults: A Randomised, Placebo-Controlled, Parallel-Groups Feasibility Study. *Front Nutr*, 9: 849902. DOI: 10.3389/fnut.2022.849902

93. Flores IO, Treviño S & Díaz A (2023). Neurotrophic Fragments as Therapeutic Alternatives to Ameliorate Brain Aging. Neural Regen Res, 18(1): 51–56.

94. Agrawal K, Chakraborty P, Dewanjee S, Arfin S, Das SS, Dey A, Moustafa M, et al. (2023). Neuropharmacological Interventions of Quercetin and Its Derivatives in Neurological and Psychological Disorders. *Neur Bio Rev*, 144: 104955. DOI: 10.1016/j.neubiorev.2022.104955

95. Dormal V, Pachikian B, Debock E, Buchet M, Copine S & Deldicque, L. (2022). Evaluation of a Dietary Supplementation Combining Protein and a Pomegranate Extract in Older People: A Safety Study. *Nutrients*, 14: 5182. DOI: 10.3390/nu14235182

96. Rode SB, Dadmal A & Salankar HV. (July 07, 2022). Nature's Gold (Moringa Oleifera): Miracle Properties. *Cureus*, 14(7): e26640. DOI: 10.7759/cureus.26640

97. Fišar Z & Hroudová, J. (2024). CoQ10 and Mitochondrial Dysfunction in Alzheimer's Disease. *Antioxidants*, 13: 191. doi: 10.3390/antiox 13020191.

98. Liang Y, Lee DYW, Zhen S, Sun H, Zhu B, Liu J, Lei D, et al. (2022). Natural Medicine HLXL Targets Multiple Pathways of Amyloid-Mediated Neuroinflammation and Immune Response in Treating Alzheimer's Disease. Phytomedicine, 104: 154158. DOI: 10.1016/j.phymed.2022.154158

99. Witte AV, Fobker M, Gellner R, Knecht S & Floel A. (2009). Caloric Restriction Improves Memory in Elderly Humans. *Proc Nati Acad Sci USA*, 106(4): 1255–1260.

100. Marioni RE, Stewart MC, Murray GD, Deary IJ, Fowkes F, Lowe G, et al. (2009). Peripheral Levels of Fibrinogen, C-Reactive Protein, and Plasma Viscosity Predict Future Cognitive Decline in Individuals without Dementia. *Psychosomatic Med*, 71(8): 901–906.

101. Calcada, D., Vianello, D., Giampieri, E., Sala, C., Castellani, G., de Graaf, A. et al. (2014). The Role of Low-Grade Inflammation and Metabolic Flexibility in Aging and Nutritional Modulation Thereof: A Systems Biology Approach. *Mech Ageing Dev*, 136–137: 138–147. DOI: 10.1016/j.mad.2014.01.004

102. Pieczynska-Zaja JM, Malinowska A, Łagowska K, Leciejewska N & Bajerska J. (2023). The Effects of Time-Restricted Eating and Ramadan Fasting on Gut Microbiota Composition: A Systematic Review of Human and Animal Studies. *Nutr Rev*, 00(0): 1–17. DOI: 10.1093/nutrit/nuad093

103. Anson, M. R., Guo, Z., de Cabo, R., Iyun, T., Rios, M., Hagepanos, A. et al. (2003). Intermittent Fasting Dissociates Beneficial Effects of Dietary Restriction on Glucose Metabolism and Neuronal Resistance to Injury from Calorie Intake. *PNAS*, 100(10): 6216–6220.
104. Rothman SM & Mattson MP. (2013). Activity-Dependent, Stress-Responsive BDNF Signaling and the Quest for Optimal Brain Health. *Neurosci*, 239: 228–240.
105. Vasconcelos AR, Yshii LM, Viel TA, Buck HS, Mattson MP, Scavone C, et al. (2014). Intermittent Fasting Attenuates Lipopolysaccharide-Induced Neuroinflammation and Memory Impairment. *J Neuroinflammation*, 11: 85.
106. Knight A, Bryan J & Murphy, C. (2016). Is the Mediterranean Diet a Feasible Approach to Preserving Cognitive Function and Reducing Risk of Dementia for Older Adults in Western Countries? New Insights and Future Directions. *Ageing Res Rev*, 25: 85–101.
107. Denis I, Potier B, Vancassel S, Heberden C & Lavialle M. (2013). Omega-3 Fatty Acids and Brain Resistance to Ageing and Stress: Body of Evidence and Possible Mechanisms. *Ageing Res Rev*, 12: 579–594. doi: 10.1016/j.arr.2013.01.007
108. Yehuda S, Rabinovitz S, Carasso RL & Mostofsky DI. (2002). The Role of Polyunsaturated Fatty Acids in Restoring the Aging Neuronal Membrane. *Neurobiol Aging*, 23: 843–853.
109. Yehuda S, Rabinovitz S & Mostofsky DI. (2005). Essential Fatty Acids and the Brain: From Infancy to Aging. *Neurobiol Aging*, 26S: S98–S102.
110. Barnard AI, Bush AI, Ceccarelli A, Cooper J, deJager CA, Erickson KI, et al. (2014). Dietary and Lifestyle Guidelines for the Prevention of Alzheimer's Disease. *Neurobiol Aging*, 35: S74–S78.
111. Choukou M-A, Olatoye F, Urbanowski R, Caon M & Monnin C. (2023). Digital Health Technology to Support Health Care Professionals and Family Caregivers Caring for Patients with Cognitive Impairment. *Scoping Rev JMIR Ment Health*, 10: e40330. DOI: 10.2196/40330
112. Giurgiu M, Ketelhut S, Kubica C, Nissen R, Doster A-K, Thron M, Timm I, et al. (2023). Assessment of 24-hour Physical Behaviour in Adults via Wearables: A Systematic Review of Validation Studies under Laboratory Conditions. *Inter J Behav Nutri Phys Activity*, 20: 68. https://doi.org/10.1186/s12966-023-01473-7

8

Sleep for the Joy of It

Sleep

Sleep for the joy of how many ways this can help you maximize your brain benefits in our Intelligent Neuroplasticity Revolution for Hacking Neuroplasticity and Your Healthy Aging Brain. Multiple factors matter.

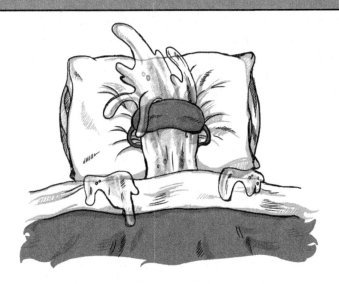

Figure 8.1
This illustration by David Yu, MD, shows our brain cell maximizing the benefits of good sleep by reducing the impact of light.

DOI: 10.4324/9781003462354-8

As the neuron sleeping in Figure 8.1 illustrates, reducing light in the night helps you sleep tight.

Many of us tend to be overworked and/or over committed in one way or another. For me, that includes serving as an expert in medical and psychiatric matters for the Involuntary Commitment (ITA) Court; being cofounder of an international project to prevent, delay onset, and/or reverse dementia; and the list goes on.

That's part of what made my heart sing when I first learned that some of our most important body work takes place while we sleep. Just reminding myself at bedtime to sleep with a lot of time in slow wave sleep for the feeding and cleaning of my brain has increased the quality of and quantity of my slumbers.

The other good news is that all is not lost when your life demands some loss of sleep. That's another reason to sleep for the *joy* of it. You are invited to use this research to sleep deeply for the benefits of nourishing and maintaining your healthy aging brain.

Set the Stage

Measuring human sleep via polysomnography[1,2] helped identify characteristics of sleep. Our sleep is noted to have periods with rapid eye movement (REM) and without (NREM). NREM sleep has subcategories of stage 1 (N1), stage 2 (N2), and stage 3 (N3). The lightest stage, N1, is a transition between being awake and being asleep. The deepest NREM sleep, N3, is also known as slow wave sleep (SWS).

A review of sleep studies[3] found that REM and NREM sleep have complementary effects in facilitating learning which involves regulating neuroplasticity in opposing directions at the synapses. Synapse formation that is specific to memories is upregulated during SWS; REM is associated with "global downregulation resulting in elimination of synapses and decreased neural firing."[4]

Pick Your Pattern

Is early to bed and early to rise, for your brain healthy and wise? Works for me! It results in my being in and out of the exercise room before most people in my time zone are awake.

Studies of patterns of sleep also considered sleep duration, number of minutes awake after sleep onset, and regularity of sleep.[5] Going to sleep later, getting fewer hours of sleep, and awakening rather than having good continuity of sleep were all sleep patterns "associated with thinner cortex and altered subcortical volumes in diverse brain regions across adolescence." Healthy sleep during childhood and adolescence could be an important intervention to promote healthy brain development.

Natural sleep patterns of 251 psychiatrically healthy participants aged 9–25 years were measured with wrist actigraphy.[6] Structural MRI studies were used to estimate brain age. "Later sleep timing (midsleep) was associated with more advanced brain aging." Less sleep occurs with going to sleep later and getting up early for school or a job. These researchers opined that this implies "late sleep timing as a prodrome or risk factor for altered brain development."

They used machine learning to identify regions of the brain with the best prediction of chronological age. Subtracting the individual's age from that indicated in the MRI findings yielded a measure referred to as a "brain age gap." They suggest research assessing whether differences in individual sleep and advanced aging of the brain in adolescents precede the beginning of "suboptimal cognitive-emotional outcomes in adolescents." Artificial intelligence with human creativity could influence the development of actigraphy and other wearables to research long-term effects of going to sleep later and getting up early for daytime obligations.

Feed and Clean with Your Glymphatic System

Slow wave sleep is when some of your most important body work is accomplished. This process might be envisioned as similar to activities in large corporate office buildings where staff leave after a great day of work. This results in hallways and offices being more open, more clear for passing through. Then, crews come in to clean and carry out waste. Shelves and refrigerators are restocked with essential nutrients and fluids to maximize staff productivity.

These are the major functions of our glymphatic system in our central nervous system which is most active while we are asleep. Sleep is essential for distribution of nutrients, growth factors, and chemical messengers between brain cells.[7] The flow of cerebrospinal fluid through and into brain tissue as well as the outflow of waste molecules, tau, and amyloid beta[8] are more active in the sleeping brain than when the person is awake.[9]

In the brain, the small blood vessels are surrounded by areas filled with fluid. Measuring these perivascular spaces is a marker of clearance of waste from the brain.[10] These perivascular spaces are involved with clearing toxins.

Diffusion tensor imaging of the perivascular space (ALPS) was used to evaluate glymphatic function in people aged 13–88 years who were without cognitive impairment.[11] "Declines in mental manipulation and short-term memory performance in the older participants were associated with a lower ALPS index and cortical atrophy in the amygdala, anterior and posterior cingulate, thalamus and middle frontal regions."

The glymphatic system, an essential facet of our neurobiological system of removing waste, depends on "astroglial water transport" crossing the blood-brain barrier.[12] Although the type and number of molecules removed by the glymphatic system from the central nervous system is still being clarified, these include potassium that can be elevated after an acute focal ischemia.[13] Molecules removed also include tau, amyloid beta, and proteins which damaged cells release.[14] Noninvasive techniques have validated these findings in our human brain.[15] With this research lacking systematic methods of assessing sleep and with most studies relying on self-report data, the use of artificial intelligence in wearable technology for gathering data on specific sleep parameters and molecular movement across the blood-brain barrier is needed to clarify how the glymphatic system works and how to personalize interventions to influence the purchases of the glymphatic system in maximizing healthy aging of your brain.

Lifestyle choices that can increase the glymphatic system power[8] include sleeping in the right lateral position to maximize the effects of gravity on the flow of fluids and blood through the brain. Additional lifestyle choices of positive influence on the glymphatic process include exercise, managing emotions associated with stress, and intermittent fasting. Also, consuming omega-3 polyunsaturated fatty acids found in marine-based fish oils[16] could mediate the function of the glymphatic system to increase amyloid-beta clearance.

That harkens back to why offering to share my tins of sardines is a gift for healthy aging of our brains. DHA for me and DHA for thee.

Photobiomodulation involves applying light of very specific wavelengths to body tissues.[17] These researchers suggest that transcranial applications of these very specific light wave lengths during sleep at night might be a "new non-pharmacological treatment" for improving quality of sleep and efficiency of the glymphatic system for better health and wellness.

The COVID-19 pandemic was among the multiple stressors having a negative impact on sleep. The dynamics of the autonomic nervous system influence and support the central nervous system integration, storage, and encoding of memory.[18] Adolescents school students in Shenzhen China were less active physically and less happy when confined to home; improving the quality of their sleep and increasing social supports was associated with their becoming happier and less sedentary.[19] More research is needed with standardized study design and methods of assessing sleep in healthy people which artificial intelligence could facilitate.

Earnings with Healthy Zzzzs

When one night of sleep deprivation in rats included gentle handling to prevent sleep, neurogenesis increased significantly initially as well as 15 and 30 days later.[20] That much increase in neurogenesis when gentle handling prevented sleep is *huge!*

Hopefully, artificial intelligence can assist in replicating this research in humans since business and other travel can significantly interfere with getting adequate sleep. On an extended flight, I apply this research finding of increased neurogenesis during sleep deprivation by using one hand to gently massage my other hand. That's just one example based on speculation. An artificial intelligence designed wearable could prompt for, measure, and capture impacts of gentle handling during extended sleep deprivation in humans.

Perhaps one of the best Earnings with Healthy *Zzzzs* is increasing neurogenesis since producing new brain cells is one of the most important ways to drive neuroplasticity in a positive direction for increasing brain power. "Brains are not computers, rather they have a reciprocal malleability of their structures which is essential for their healthy function."[21] Neurogenesis in the hippocampus plays a key role in neurocognitive functions, including memory. These new brain cells add new synapses, new connections, and life-course plasticity and function.

Although neurogenesis continues in the hippocampus across the human lifespan, adult neurogenesis decreases rapidly in Alzheimer's disease which might be a contributing factor to memory and learning impairments in individuals with this disease.[22] In their elegant overview of recent studies of hippocampal neurogenesis in adults, they consider the impacts of impaired adult neurogenesis in learning and memory impairments as seen in Alzheimer's disease and raise questions about activating neurogenesis in patients with Alzheimer's for therapeutic potential.

Looking at data from 37,533 healthy people in the UK Biobank, researchers noted that the best cognitive performance was observed in those who slept seven hours each day.[23] To assess cognitive performance, these researchers looked at executive functioning estimated from tasks of speed of processing or working memory.

It is noteworthy that, although these neurocognitive skills tend to decline with age, these findings of *best executive functioning with 7 hours of sleep included the 12,006 individuals who were more than 60 years old.* That is also about a third of a very large sample of healthy individuals.

People who slept six-to-eight hours daily "had significantly greater grey matter volume in 46 of 139 different brain regions including the orbitofrontal cortex, hippocampi, precentral gyrus, right frontal pole and cerebellar subfields." Thus, they opined that their "findings highlight the important relationship between the modifiable lifestyle factor of sleep duration and cognition as well as a widespread association between sleep and structural brain health."

That's another reason my exercise routine is accomplished first thing out of bed. After about seven hours of sleep, with as much short-wave sleep as possible, time exercising evaporates.

Data from the UK Biobank of 14,206 participants looked at brain scans as well as measures of sleep and cognition.[24] "Every additional hour of sleep above 7 h/day was associated with 0.10–0.25% lower brain volumes." Sleeping more than nine hours and sleeping less than six hours were both associated with "lower brain volumes and cognitive measures (memory, reaction time, fluid intelligence)." Dozing during the daytime was also "associated with lower brain volumes (grey matter and left hippocampus volume) and lower cognitive measures (reaction time and fluid intelligence)."

Similar brain microstructure preservation with quality sleep 6.3 years prior to MRI was found in 146 elders who were dementia free.[25] They opined that "Optimizing sleep behaviors throughout the life-course may help to preserve healthy brain aging."

By paring sounds with material learned when awake and presenting these stimuli during REM sleep, measured brain reactivation strength in human REM predicted overnight improvement in task performance when awake.[26] In a small sample with 34 young healthy adults, neuroplastic impacts of sleep and sleep deprivation were assessed. The first day they had noninvasive brain studies done before and after given 40 minutes of learning to navigate in a virtual environment where they were required to reach targets as fast as possible.[27] Then, they had a night of either normal sleep or were sleep deprived. After nights of recovery sleep, they were scanned again before and after the navigational task. The researchers described "sleep-related remodeling of neurites and glial cells subtending learning and memory processes in basal ganglia and hippocampal structures."

Sleep is one of the critical regulators of memory.[28] These memory benefits of sleep include vocabulary, associative memory, and declarative memories; sleep is also beneficial in retaining sensory information and motor skills.[29]

Using a network of populations of thalamic and cortical neurons, new learning memory traces were formed in the wake state.[30] Given a replay during the slow wave sleep state "increased synaptic connectivity" enabling "indirect memory recall." They opined that their study predicts sleep can strengthen and form relational memories.

Reactivation during sleep of specific training in counter-social-bias enhanced training effects.[31] This was a small study with 28 young adults.

A study of 6,050 adults aged 65 or older[32] found greater quality of life and independent functioning in individuals who had adequate sleep as compared to those who reported insomnia. One of the neuroprotective mechanisms of adequate sleep may be its reduction of inflammation that can be associated with aging[33] as well as with decreased neurogenesis as observed in animal models.[34] Factors improving sleep efficiency and quality include being physically active according to a review of studies reporting on 2,612 patients.

Poor Snooze Zzzzs and You Could Lose

Chronic sleep deprivation in animals resulted in increased inflammatory molecules and decreased BDNF which is crucial to many components of neuroplasticity.[35] A recent review of animal research[36] concludes that the major physiological challenge created by sleep deprivation can include "cognitive deficits, inflammation, general impairment of protein translation, metabolic imbalance, and thermal deregulation." Sleep is essential for removal of waste and distribution of "glucose, lipids, amino acids, growth factors, and neuromodulators."[7]

A study using data from the UK Biobank queried 36,468 people whose average age was 63.6.[37] These researchers found that sleep-related symptoms such as snoring and hypersomnolence are risk factors for cognitive impairment in elders. Their fluid intelligence scores were lower; reaction times were slower; and executive functions as well as memory scores were lower with self-reported sleep-related symptoms. That's a lot to lose with poor snooze.

While the best executive functioning as reflected in speed of processing and working memory is associated with daily sleep of seven hours, this "decreased for every hour below and above this sleep duration."[23] Although age is considered the strongest predictor of decline in neurocognition, sleep is one of the modifiable lifestyle factors which could help preserve brain function and structure. Negative brain white matter changes were found on MRI with dementia-free elders reporting worse quality of sleep over the recent 6.3 years.[25]

Sleep fragmentation and sleep of poor quality are associated with accelerated aging by as much as two years above chronological age.[38] These modifiable factors which increase the risk of dementia and cognitive decline "should no longer be viewed as intrinsically related to normal ageing but" should be seen as factors we can modify to influence healthy brain aging rather than lose it.

A review of studies that included 577,932 participants from Asian and North American upper-middle-income countries found that being exposed to light at night impaired sleep.[39] This impairment was increased 22% for people with higher levels of exposure to light at night; thus, indoor exposure to light at night is another modifiable factor to protect sleep. Despite the large number of participants in this review, the authors caution that the findings may not generalize beyond the populations studied.

Data on 22,599 participants in the National Health and Nutrition Examination Survey (NHANES) database indicated that sleep disturbance can be reflected in inflammatory biomarkers associated with being sedentary.[40] They opine that the relationship between sleep, exercise, and sedentary behavior suggests that sleep and inflammatory biomarkers can be effectively improved with exercise.

Sleep, gut microbiota, and brain interactions are important as early as during infancy. Behavioral impacts of the "sleep-gut-brain linkages" have been noted in three-month-old children[41] with a "relationship between infant daytime sleep and alpha diversity of gut microbiota." A review of studies[42] described sleep and gut microbiota relationships as bidirectional and opined that influencing gut microbiota could improve sleep disorders. However, this field of study is in its infancy.

The influence on neuroplasticity during adolescent development is an important area of study. Whole brain volume and thinning of the volume in the cortical area were positively and significantly associated with sleep duration and with poorer sleep during adolescence.[5]

Nursing staff working night shift have the occupational hazard of changes in brain function.[43] Twenty nurses tested while rested had reduced memory when sleep deprived which was associated with changes in their fMRI.

After being deprived of sleep for 36 hours, male college students had a significant decrease in accuracy of their memory.[44] After they had recovery sleep of eight hours, there was some improvement in their working memory.

An overnight study with 92 young healthy adults compared four situations: sleep restriction to 5–6 hours; average sleep of 8–9 hours; and high intensity interval training before both average and restricted sleep.[45] Memory for 80 face-name pairs was tested in the evening for immediate recall and again in the morning after sleep for delayed recall. Acute high intensity interval training in the evening prior to sleep restriction reduced the negative effects on long-term declarative memory.

Another essential role of sleep may be to restore brain energy metabolism since wakefulness consumes more energy particularly in grey brain matter.[46] Mice showed decreased function in the blood-brain barrier with sleep deprivation.[47] As described above, this interface between circulation and the brain is crucial to adequate supply of nutrients, growth factors, and oxygen to brain cells as well as removing molecules, including amyloid beta, tau, and waste. Sleep deprivation in mice resulted in neuroinflammation in the hippocampus and associated deficits in learning and memory.[48] In a population of 2,822 men aged 67 and older, measured and reported sleep disturbance was associated with cognitive decline.[49]

A review of studies[34] found that accumulated sleep deprivation and sleep fragmentation greater than 24 hours was associated with a decrease in neurogenesis that was not quickly reversible. Memory deficits and mood effects noted with sleep-deprived humans may have some association with impaired neurogenesis.[50]

For me, that again raises the question. If humans have gentle handling with sleep deprivation of a day, can they have increased neurogenesis initially as well as 15 and 30 days later as found by Zucconi and colleagues?[20] Artificial intelligence could assist in researching this in humans.

Seventeen people who had never complied with treatment of chronic obstructive sleep apnea showed "diffuse reduction" in white brain matter integrity that was associated with cognitive dysfunction as measured with neuropsychological testing.[51] However, after one year of compliance with treatment, brain pathology had improved significantly along with "significant improvements involving memory, attention, and executive functioning."

After 48 depressed patients experienced total sleep deprivation for a day, their mood improved and they ruminated less.[52] White matter (WM) structural changes measured by diffusion tensor imaging used statistical analysis of fractional anisotropy (FA). Depressed patients "showed significant ($p < 0.05$, corrected) decreases in FA values in multiple WM tracts, including the body of the corpus callosum and anterior corona radiata" after total sleep deprivation. These changes were not noted in healthy individuals after total sleep deprivation. These researchers opined that "WM tracts including the superior corona radiata and posterior thalamic radiation could be potential biomarkers of the rapid therapeutic effects of TSD. Changes in superior corona radiata FA, in particular, may relate to improvements in maladaptive rumination."

Artificial intelligence and machine learning using wearable devices can assist clinicians in diagnosing, treating, and monitoring severity of sleep apnea,[53] depression, and progress in treatment of these and other illnesses that are associated with sleep problems. It can provide measurements of activities in healthy individuals that are associated with sleep benefits. Artificial intelligence can also help refine and broaden the reach of research in studying characteristics of healthy sleep that are associated with driving neuroplasticity in a positive direction. Further development of these wearable technologies could assist in reminding individuals about recommendations for improved sleep as well as providing positive reinforcement for compliance and progress.

Summary for Zzzzs to Maximize Neurogenesis and Neurocognition

As much time as I spend on important tasks, it is delicious to learn that I can sleep through some other very important work. Feeding and cleaning my brain.

We can celebrate SLEEP for the JOY of It because some of our most important brain and body work takes place while we sleep. Feeding our hungry brain to restore brain energy metabolism *and* removing waste molecules is done best while in sleep we rest. We can increase the power of this glymphatic system with exercise, managing emotions associated with stress, intermittent fasting, and consuming omega-3 polyunsaturated fatty acids.

Photobiomodulation might be a "new non-pharmacological" approach to improve sleep processes in our future.[17] Further research is needed on this process.

It is significant that executive functioning, as estimated from tasks of speed of processing and working memory, were best with seven hours of sleep. It is empowering and encouraging that this included 12,006 individuals who were more than 60 years old. Sleeping 6–8 hours daily preserved brain structural health. Slow wave sleep regulates neuroplasticity by increasing synapse formation that is specific to memories.

Neurogenesis is one of the most important ways to drive neuroplasticity in a positive direction for enhancing neurocognition. Neurogenesis in the hippocampus plays a key role in neurocognitive functions, including memory. These new brain cells add new synapses, new connections, and life-course plasticity and function. It is noteworthy that, when gentle handling of rats prevented sleep for one night, their neurogenesis increased significantly initially as well as 15 and 30 days later.[20] When an overnight flight or work shift is required, can humans reap similar benefits by, for example, using their one hand to gently massage the other hand? The jury is still out.

Artificial intelligence could help develop wearables to assess for this as well as assessing for the impact of exercise prior to and after the loss of sleep. Sleep fragmentation and sleep of poor quality which have been associated with accelerated aging are modifiable factors which can be associated with neuroinflammation which has a negative effect on neuroplasticity. Sleep and inflammation can be effectively improved by exercise.

Artificial intelligence and machine learning can assist clinicians and individuals in diagnosing, treating, and monitoring progress in personalized evidence-based interventions to improve sleep for driving neuroplasticity in a positive direction to maximize neurocognition as well as healthy brain structure. The importance of measuring, modifying, and giving positive reinforcement for lifestyle factors of quality and quantity of sleep cannot be overstated.

Artificial intelligence could increase our database on sleep architecture, function, and various benefits. It can complement an individual's perceptions of factors of their sleep which research has shown to be important; for example, total time asleep as reported by participants explained 61% of the variance.[54] Sleep EEG for detecting dementia in 8,044 participants was effective in discriminating cognitively normal individuals from those with dementia as determined by tests as well as clinical diagnosis.[55] We can proceed with a Positive Psychology perspective[56] and maximize our body work as we sleep for the healthy aging of our brains.[57]

Artificial intelligence with human creativity, problem solving, and guidance can contribute to designing wearables, nearables, and research of specific characteristics of sleep that contribute to maximizing status of functioning, diagnosis, and best positive neuroplastic impact on brain chemistry, architecture, and function. This could increase our ability to SLEEP for the JOY of appreciating how many ways we are hacking neuroplasticity for our healthy aging brains.

References

1. Iber C, Ancoli-Israel S, Chesson C & Quan SF. (2007). *The AASM Manual for the Scoring of Sleep and Associated Events*. Westchester: American Academy of Sleep Medicine.
2. Liu M, Zhu H, Tang J, Chen H, Chen C, Luo J, & Chen W. (2023). Overview of a Sleep Monitoring Protocol for a Large Natural Population. *Phenomics,* 3: 421–438. https://doi.org/10.1007/s43657-023-00102-4
3. Uji M & Tamaki M. (2023). Sleep, Learning, and Memory in Human Research Using Noninvasive Neuroimaging Techniques. *Neurosci Res*, 189: 66–74. https://doi.org/10.1016/j.neures.2022.12.013
4. Niethard N, Burgalossi A & Born J, (2017). Plasticity During Sleep Is Linked to Specific Regulation of Cortical Circuit Activity. *Front Neural Circuits*, 11. https://doi.org/10.3389/fncir.2017.00065
5. Jalbrzikowski M, Hayes RA, Scully KE, Franzen PL, Hasler BP, Siegle GJ, Buysse DJ, Dahl RE, Forbes EE, Ladouceur CD, et al. (2021). Associations Between Brain Structure and Sleep Patterns Across Adolescent Development. *Sleep*, 44(10): zsab120.
6. Soehner AM, Hayes RA, Franzen PL, Goldstein TR, Hasler BP, Buysse DJ, et al. (2023). Naturalistic Sleep Patterns Are Linked to Global Structural Brain Aging in Adolescence. *J Adolescent Health*, 72: 96e104. https://doi.org/10.1016/j.jadohealth.2022.08.022
7. Jessen NA., Munk ASF, Lundgaard I & Nedergaard M (2015). The Glymphatic System: A Beginner's Guide. *Neurochem Res*, 40(12): 2583–2599.
8. Reddy OC & van der Werf YD. (2020). Harnessing the Power of the Glymphatic System Through Lifestyle Choices. *Brain Sci*, 11: 868.
9. Xie L, Kang H, Xu Q, Chen MJ, Liao Y, Thiyagarajan M, O'Donnell J, Christensen DJ, Nicholson C, Iliff JJ, et al. (2013). Sleep Drives Metabolite Clearance from the Adult Brain. *Science*, 342: 373–377. https://doi.org/10.1126/ science.1241224
10. Aribisala BS, Hernandez MDCV, Okely JA, Cos SR, Ballerini L, Dickie DA, Wiseman SJ, et al. (2023). Sleep Quality, Perivascular Spaces and Brain Health Markers in Ageing - A Longitudinal Study in the Lothian Birth Cohort 1936. *Sleep Med*, 106: 123e131. https://doi.org/10.1016/j.sleep.2023.03.016
11. Hsiao W-C, Chang H-I, Hsu S-W, Lee C-C, Huang s-H, Cheng C-H, Huang C-W & Chang C-C. (2023). Association of Cognition and Brain Reserve in Aging and Glymphatic Function Using Diffusion Tensor Image-along the Perivascular Space (DTI-ALPS). *Neuroscience*, 524: 11–20. https://doi.org/10.1016/j.neuroscience.2023.04.004. Online ahead of print.
12. Iliff JJ, Wang M, Zeppenfeld DM, Venkataraman A, Plog BA, Liao Y, Deane R & Nedergaard M. (2013). Cerebral Arterial Pulsation Drives Paravascular CSF–Interstitial Fluid Exchange in the Murine Brain. *J Neurosci*, 33(46): 18190. https://doi.org/10.1523/JNEUROSCI.1592-13.2013
13. Monai H, Want X, Yahagi K, Lou N, Mestre H, Xu Q, Abe Y, Yasui M, Iwai Y, Nedergaard, M & Hirase H. (2019). Adrenergic Receptor Antagonism Induces Neuroprotection and Facilitates Recovery from Acute Ischemic Stroke. *Proc Natl Acad Sci USA*, 116: 11010–11019. DOI: 10.1073/pnas.1817347116

14. Bohr T, Hjorth PG, Holst SC, Hrabetova S, Kiviniemi V, Lilius T, Lundgaard I, Mardal K-A, et al. (2022). The Glymphatic System: Current Understanding and Modeling. *iScience*, 29: 104987. https://creativecommons.org/licenses/by-nc-nd/4.0/

15. Eide PK, Vinje V, Pripp AH, Mardal KA & Ringstad G. (2021). Sleep Deprivation Impairs Molecular Clearance from the Human Brain. *Brain*, 144: 863–874. https://doi.org/10.1093/brain/awaa443

16. Ren H, Luo C, Feng Y, Yao X, Shi Z, Liang F, Kang JX, Wan J-B, Pei Z & Su H. (2017). Omega-3 Polyunsaturated Fatty Acids Promote Amyloid-B Clearance from the Brain Through Mediating the Function of the Glymphatic System. *FASEB J*, 31: 282–293. www.fasebj.org

17. Valverde A, Hamilton C, Moro C, Billeres M, Magistretti P & Mitrofanis J (2023) Lights at Night: Does Photobiomodulation Improve Sleep? *Neural Regen Res*, 18(3): 474–477.

18. Whitehurst LN, Subramoniam A, Krystal A & Prather AA. (2022). Links between the Brain and Body during Sleep: Implications for Memory Processing. *Trends Neurosci*, 45(3). https://doi.org/10.1016/j.tins.2021.12.007

19. Zou L, Wang T, Herold F, Ludyga S, Liu A, Zhang Y, Healy S, Zhang Z, Kuang J, Taylor A, Kramer AF, Chen S, Tremblay MS & Hossain M. (2023). Associations between Sedentary Behavior and Negative Emotions in Adolescents during Home Confinement: Mediating Role of Social Support and Sleep Quality. *Int J Clin Health Psychol*, 23: 100337. https://doi.org/10.1016/j.ijchp.2022.100337.

20. Zucconi GG, Cipriani S, Balgkouranidou I & Scattoni R. (2006). "One Night" Sleep Deprivation Stimulates Hippocampal Neurogenesis. *Brain Res. Bull*, 69(4): 375–381.

21 Kempermann G, Song H & Gage FH. (2023). Adult Neurogenesis in the Hippocampus. *Hippocampus*, 33: 269–270. https://doi.org/10.1002/hipo.23525

22. Choi SH & Tanzi RE. (2023). Adult Neurogenesis in Alzheimer's Disease. *Hippocampus*, 33: 307–321. https://doi.org/10.1002/hipo.23504

23. Tai XY, Chen C, Manohar S & Hussain M. (2022). Impact of Sleep Duration on Executive Function and Brain Structure. *Commun Biol*, 5: 201. https://doi.org/10.1038/s42003-022-03123-3

24. Namsrai T, Ambikairajah A & Cherbuin N. (2023). Poorer Sleep Impairs Brain Health at Midlife. *Sci Rep*, 13: 1874. https://doi.org/10.1038/s41598-023-27913-9

25. Tsiknia AA, Parada H, Banks SJ & Reas ET. (2023). Sleep Quality and Sleep Duration Predict Brain Microstructure among Community-Dwelling Older Adults. *Neurobiol Aging*, 125: 90–97. https://doi.org/10.1016/j.neurobiolaging.2023.02.001

26. Abdellahi MEA, Koopman ACM, Treder MS & Lewis PA. (2023). Targeted Memory Reactivation in Human REM Sleep Elicits Detectable Reactivation. *Elife*, 12: e84324. https://doi.org/10.7554/eLife.84324

27. Villemonteix T, Guerreri M, Deantoni M, Balteau E, Schmidt C, Stee W, Zhang H & Peigneux P. (2023). Sleep-Dependent Structural Neuroplasticity after a Spatial Navigation Task: A Diffusion Imaging Study. *J Neurosci Res*, 101: 1031–1043. https://doi.org/10.1002/jnr.25176.

28. Zhou Z & Norimoto H. (2023). Sleep Sharp Wave Ripple and Its Functions in Memory and Synaptic Plasticity. *Neurosci Res*, 189: 20–28. https://doi.org/10.1016/j.neures.2023.01.011

29. Rothschild, G (2019). The Transformation of Multi-sensory Experiences into Memories during Sleep. *Neurobiol Learn Mem*, 160: 58–66. https://doi.org/10.1016/j.nlm.2018.03.019.

30. Tadros T & Bazhenov M (2022). Role of Sleep in Formation of Relational Associative Memory. J Neuroscience, 42(27): 5330–5345.

31. Xia T, Antony J W, Pakker KA & Hu X. (2023). Targeted Memory Reactivation during Sleep Influences Social Bias as a Function of Slow-Oscillation Phase and Delta Power. *Psychophysiology*, 60: e14224. https://doi.org/10.1111/psyp.14224

32. Spira AP, Kaufmann CN, Kasper JD, Ohayon MM, Rebok GW, Skidmore E, et al. (2014). Association between Insomnia Symptoms and Functional Status in US Older Adults. *Psychol Sci Soc Sci*, 69(7): S35–S41, https://doi.org/10.1093/geronb/gbu116

33. Irwin M. (2014) Sleep and Inflammation in Resilient Aging. *Interface Focus*, 4(5): 20140009.

34. Guzman-Marin R & McGinty D. (2006). Sleep Deprivation Suppresses Adult Neurogenesis: Clues to the Role of Sleep in Brain Plasticity. *Sleep Biol Rhythms*, 4: 27–34.

35. Zielinski MR, Kim Y, Karpova SA, McCarley RW, Strecker RE & Gerashchenko D. (2014). Chronic Sleep Restriction Elevates Brain Interleukin-1 Beta and Tumor Necrosis Factor-Alpha and Attenuates Brain-Derived Neurotrophic Factor Expression. *Neurosci Lett*, 580: 27–31. https://doi.org/10.1016/j.neulet.2014.07.043

36. da Costa Souza A & Ribeiro S. (2015). Sleep Deprivation and Gene Expression. Current Topics Behav Neuroscience. Springer-Verlag Berlin Heidelberg. https://doi.org/10.1007/7854_2014_360.

37. Yu J, Morys F, Dagher A, Lajoie A, Gomes T, Ock EY, Kimoff J & Kaminska M. (2023). Associations Between Sleep-Related Symptoms, Obesity, Cardiometabolic Conditions, Brain Structural Alterations and Cognition in the UK Biobank. *Sleep Med*, 103: 41e50. https://doi.org/10.1016/j.sleep.2023.01.023

38. Ramduny J, Bastiani M, Huedepohl R, et al (2022). The Association Between Inadequate Sleep and Accelerated Brain Ageing. *Neurobiol Aging*, 114: 1–14.

39. Xu Y-X, Zhang J-H, Tao, F-B & Sun Y. (2023). Association Between Exposure to Light at Night (LAN) and Sleep Problems: A Systematic Review and Meta-analysis of Observational Studies. *Sci Total Environ*, 857: 159303. https://dx.doi.org/10.1016/j.scitotenv.2022.159303

40. You Y, Chen Y, Fang W, Li X, Wang R, Liu J & Ma X. (2023). The Association Between Sedentary Behavior, Exercise, and Sleep Disturbance: A Mediation Analysis of Inflammatory Biomarkers. *Front Immunol*, 13: 1080782. https://dx.doi.org/10.3389/fimmu.2022.1080782

41. Schoch SF, Castro-Mejia JL, Krych L, Leng B, Kot W, Kohler M, et al. (2022). From Alpha Diversity to Zzz: Interactions Among Sleep, the Brain, and Gut Microbiota in the First Year of Life. *Prog Neurobiol*, 209: 102208. https://doi.org/10.1016/j.pneurobio.2021.102208

42. Wang Z, Wang Z, Lu T, Chen W, Yan W Yuan K, Shi L, Liu X, Zhou X, Shi J, Vitiello MV, Han Y & Lu L. (2022). The Microbiota-Gut-Brain Axis in Sleep Disorders, *Sleep Med Rev*. 65: 101691. https://doi.org/10.1016/j.smrv.2022.101691. [Epub 2022 August 31].

43. Yan F-X, Lin J-L, Lin J-H & Lin Y-J. (2023). Altered Dynamic Brain Activity and Its Association with Memory Decline after Night Shift-Related Sleep Deprivation in Nurses. *J Clin Nurs*, 32: 3852–3862. https://doi.org/10.1111/jocn.16515

44. Peng Z, Hou Y, Xu L, Wang H, Wu S, Song T, Shao Y & Yang Y (2023) Recovery Sleep Attenuates Impairments in Working Memory Following Total Sleep Deprivation. *Front Neurosci*, 17: 1056788. https://doi.org/10.3389/fnins.2023.1056788

45. Frimpong E, Mograss M, Zvionow T, Paez A, Aubertin-Leheudre M, Bherer L, et al. (2023). Acute Evening High-Intensity Interval Training May Attenuate the Detrimental Effects of Sleep Restriction on Long-Term Declarative Memory. *Sleep*, 46(7): zsad119. https://doi.org/10.1093/sleep/zsad119

46. Plante DT, Trksak GH, Jensen E, Penetar DM, Ravichandran C, Riedner BA, et al. (2014) Gray Matter-Specific Changes in Brain Bioenergetics after Acute Sleep Deprivation: A 31P Magnetic Resonance Spectroscopy Study at 4 Tesla. *Sleep*, 37(12): 1919–1927. https://doi.org/10.5665/sleep.4242

47. He SJ, Hsuchou H, He Y, Kastin AJ, Wang & Pan W. (2014). Sleep Restriction Impairs Blood–Brain Barrier Function. *J Neuroscience*, 34(44): 14697–14706. https://doi.org/10.1523/JNEUROSCI.2111–14.2014

48. Zhu B, Dong Y, Xu Z, Gompf HS, Ward SA, Xue Z, et al. (2012). Sleep Disturbance Induces Neuroinflammation and Impairment of Learning and Memory. *Neurobio Dis*, 48(3): 348–355. https://doi.org/10.1016/j.nbd.2012.06.022.

49. Blackwell T, Yaffe K, Laffan A, Ancoli-Israel S, Redline W, Ensrud KE, et al. (2014). Associations of Objectively and Subjectively Measured Sleep Quality with Subsequent Cognitive Decline in Older Community-Dwelling Men: The MrOS Sleep Study. *Sleep*, 37(4): 655-63. DOI: 10.5665/sleep.3562.

50. Mueller AD, Meerlo P, McGinty D & Mistlberger RE. (2013). Sleep and Adult Neurogenesis: Implications for Cognition and Mood. *Curr Top Behav Neurosci* https://dx.doi.org/10.1007/7854_2013_251

51. Castronovo V, Scifo P, Castellano A, et al. (2014). White Matter Integrity in Obstructive Sleep Apnea Before and After Treatment. *Sleep*, 37(9). https://dx.doi.org/10.5665/sleep.3994

52. Taraku B, Zavaliangos-Petropulu A, Loureiro JR, Al-Sharif NB, Kubicki A, Joshi SH, Woods RP, et al. (2023) White Matter Microstructural Perturbations After Total Sleep Deprivation in Depression. *Front Psychiatry*, 14: 1195763. https://dx.doi.org/10.3389/fpsyt.2023.1195763

53. Bazoukis G, Bollepalli SC, Chung CT, Li X, Tse G, Bartley BL, Batool-Anwar S, Quan SF & Armoundas AA. (2023). Application of Artificial Intelligence in the Diagnosis of Sleep Apnea. *J Clin Sleep Med*, 19(7): 1337–1363. https://dx.doi.org/10.5664/jcsm.10532

54. Yeh A-Y, Pressler SJ & Giordani BJ. (2023). Actigraphic and Self-reported Sleep Measures in Older Adults: Factor Analytic Study. *West J Nurs Res*, 45(1): 4–13. DOI: 10.1177/01939459211037054

55. Ye EM, Sun H, Krishnamurthy PV, Adra N, Ganglberger W, Thomas RJ, Lam AD, et al (2023). Dementia Detection from Brain Activity during Sleep. *Sleep*, 46(3): zsac286. https://doi.org/10.1093/sleep/zsac286
56. Peterson, C. (2008) What Is Positive Psychology, and What Is It Not? Psychol Today, posted online May 16, 2008; https://www.psychologytoday.com/us/blog/the-good-life/200805/what-is-positive-psychology-and-what-is-it-not? (Accessed online July 2, 2023).
57. Voumvourakis KI, Sideri E, Papadimitropoulos GN, Tsantzali I, Hewlett P, Kitsos D, Stefanou M, Bonakis A, Giannopoulos S, Tsivgoulis G, et al. (2023) The Dynamic Relationship between the Glymphatic System, Aging, Memory, and Sleep. Biomedicines, 11: 2092. https://doi.org/ 10.3390/ biomedicines11082092

9

AI to the Smorgasbord to Hasten and Refine Our Intelligent Neuroplasticity Revolution

~

Figure 9.1
This illustration by David Yu, MD, captures one of my experiences. While on a business trip and in the hotel exercise room, I was studying neuroplasticity, taking notes, and heard the comment from the men behind me. "Hey! Check out that gal. She's got it all going on."

DOI: 10.4324/9781003462354-9

The words in Figure 9.1 were overheard in the exercise room: That comment came from two young men on the elliptical machines behind me.

All things are ready if your mind be so.

William Shakespeare

Our table is full of evidence-based interventions for FUN furthering Flynn,[1,2] the increase in human intelligence in recent decades.[1] We can, and we must, enhance our brain power, brain health, and quality of life *at any age* in ways that *also* could improve our physical health, our well-being, our prosociality, and our effective work to make our world a better place for *ALL*. Simultaneously using more than one of these choices from our smorgasbord of evidence-based choices can yield even greater positive influence on our neuroplasticity for *Enriching Heredity*.[3,4]

That is the message in the illustration of me above. It captures the (1) benefits of aerobic exercise, which includes increased neurogenesis, plus (2) the neurobics of reading and writing on topics that:

a. will increase the percentage of these new brain cells that survive because new neurons are young and excitable requiring stimulation as an invitation to survive and get integrated into an already very complex brain AND

b. reading the subject of greatest personal interest while being aerobic will influence the place in the brain that the new neurons will integrate in response to the topic studied.

That's another way of saying that the subject being studied will influence the career choice of the new brain cells. What you study might tell your new brain cells where to go.

My neurobic was studying neuroplasticity while aerobic. What's your form of FUN?

Hacking neuroplasticity for promoting healthy aging, best possible brain health, enhanced brain power, and prosociality *at any age* is urgent. Artificial intelligence used with human creativity, imagination, reasoning, and directions can help us do that.

How likely is it that our Intelligent Neuroplasticity Revolution can promote higher levels of human intelligence than we can imagine? With artificial intelligence, "the future is not what it used to be."[5] We can hack neuroplasticity for our healthy-aging brains with choices from an abundant smorgasbord of evidence-based interventions *and* we can do this in ways that also increase physical health and prosocial behaviors which could help increase our capacity for peaceful coexistence.

Any sufficiently advanced technology is indistinguishable from magic.

Isaac Asimov (1920–1992)
Author of classic works of science fiction.
Professor of Biochemistry

Progress We Have Enjoyed

Positive emotions promote healthy aging. We have enjoyed much progress.

Flynn Effect[1,2]

The Flynn effect suggests that we must be doing some things right since increases in human intelligence have been documented in recent decades in 34 countries.[1] Using artificial intelligence and human creativity in applying options available on our smorgasbord of evidence-based interventions, we can work together to increase progress and do so for *more* of our global citizens.

SuperAgers among Us

The tsunami of research reporting on SuperAgers is valuable to help us design our strategic and personally tailored hacking of neuroplasticity. Structural and functional healthy brain aging were found in SuperAgers even with amyloid deposits, suggesting that brain resilience could be protective despite neurodegeneration.[6,7] Brain network functional connectivity studied with fMRI could identify SuperAgers.[8]

Positive-Agers, Too

In addition to SuperAgers, 59.7% of the 1,303 individuals in the UK Biobank sample were identified as positive agers in a study using machine learning.[9] They biochemically distinguished adults showing neurocognitive gains rather than decline across time. It is significant that "keeping fasting serum glucose levels below 3.2 mmol/L was associated with the best chances of" healthy brain and cognitive aging as noted in this machine learning look at 1,303 participants in the UK Biobank sample.

Centenarians and Supercentenarian Role Models

The 100-plus Study included 332 Dutch centenarians who were cognitively healthy,[10] were from higher economic status, completed more education, and had more children than others their age. Those with greater cognitive capacity lived longer.

Of 27 centenarian brains, there were three "supernormal" centenarians who had retained good mental functioning; in their neuropathological findings "no apparent senile changes or ischemic lesions" were found. The postmortem findings of the 27 brains were "not fundamentally different" than findings of brains of less elderly.[11]

In 57 dementia-free centenarians, the finding that "stronger functional connectivity between right frontoparietal control network," and a stronger functional connectivity compared to subjects aged 76–79, demonstrated a more intact bilateral neuronal efficiency.[12]

Several senior successes in centenarians and supercentenarians add to what we might want to factor into our efforts. Howard Tucker, MD, was listed in the GuinnessWorldRecords.com as the oldest physician in practice, having been a "practicing doctor and neurologist for more than seven decades" and he is still seeing patients at the age of 101 years. Robert Marchand[13] set three world records for track bicycling. Marchand also gave us metric evidence of improving cardiovascular functioning in a centenarian, the *first* such evidence in a centenarian!! AND his was a 13% improvement in cardiovascular functioning at age 103!!

Artificial intelligence can help us design measurement tools that are specific to the centenarians and supercentenarians so that we (1) will learn more from the wisdom of the elders and (2) can develop interventions that have the goal of assisting our oldest old in maximizing their superior senior successes with a healthspan that approximates their lifespan. With joint and strategic efforts, might our centenarians and supercentenarians surprise us with how much more the human intellect can achieve? Is this likely to be greater than we could imagine? Artificial intelligence with human oversight can help design research and evidence-based interventions to address these questions.

Decreased Dementia Incidence

Research showing a decreased rate of incidence of dementia by 13% per calendar decade in Europe and in the USA is also encouraging. Healthy aging without dementia is achievable[14–17] even with lesions postmortem.[11]

Although percentages varied, there were centenarians with no dementia in many countries: 37% in Denmark[14]; 54% in Italy[18]; and 21% in New

England.[19] The Fordham Centenarian Study found no, or very few, cognitive limitations in 119 centenarians.[16] The Framingham Heart Study[20] found a decline in incidence of dementia across three decades.

Rx to Prevent, Delay Onset, and/or Reverse Dementia

Neuroscience provides evidence-based interventions to "prevent, delay onset, and/or reverse" cognitive decline and dementia[21-23] and to enhance brain architecture and function.[3,24-38]

Studies suggest that senile dementia can be seen as "age-related" because it occurs within specific ages, "rather than as an 'age-related' disorder (that is, caused by the aging process itself)."[39] Dutch centenarians who scored 26 or above on the MMSE maintained their high level of cognitive performance two years later.[40,41]

Among the reasons exercise remains important is it increases brain factors which might play a role in "reversing brain aging" and "improve cognitive functioning."[42-44] Chinese elders with MCI appreciated cognitive gains, less depression, and better balance after 12 weeks of square dancing.[45] After 18 months of dancing, participants had improved balance as well as increased volume of the hippocampus[46] as measured by MRI as compared to individuals whose physical fitness routine was conventional. Increased plasma BDNF was found in dancers whose attention and spatial memory were improved.[47]

Evolving neuroscience is increasingly empowering, encouraging, and easily applied. Since people prefer to do what is easy, portable, fun, culturally adaptable, and free or low cost, it is fortunate that evolving neuroscience research is fleshing out the many neurocognitive and other healthy-aging benefits that music and dance influence.

Noninvasive tests indicate that newborn babies have memory for music they heard while they were in utero. Premature infants stabilized faster and required less time in the NICU when they heard soothing music. Youth learn better and are more prosocial with music in their lives. Elderly individuals with dementia choked less, were more present, and improved with the sound of music.

People hearing music tend to move which is far preferable to the usual sedentary status. Some respond to music by dancing which, if at an aerobic pace, yields even more brain and general healthy-aging benefits. If dancing with someone, the addition of social interaction is another bonus.

Then, there's the potential for huge benefits from touch. Recall the life-saving impact captured in the photo of the Rescuing Hug where the arm of the healthier premature infant around the shoulders of her weaker twin sister who was at risk of death helped the fragile premature twin survive and thrive.

Also empowering is the research of Dr. Marian Diamond,[3] the mother of neuroplasticity. The solitary research protocol change, which she labeled "TLC," of having the laboratory technicians remove the rats from the cage, hold them, and talk to them resulted in a 50% increase in their lifespan to the equivalent of 90 human years AND her animals continued to increase the complexity of their brain cells for "Enriching Heredity" to their equivalent of 90 human years!! That might also mean that Dr. Marian Diamond was the mother of Positive Psychology. Her videos include her brief description of her research and include her opinion that the human brain "can stay healthy and active for a full 100 years."

Marian Diamond was prescient in predicting that "Enriching Heredity" across a vigorous longevity to 100 years of age would also be achievable in humans. Subsequent research has found this to be the case.

That's a good segue back to Music and Dance which can include holding, talking, touch, "TLC," social interaction, some level of exercise which might be aerobic, complex new learning for some of us, the need to coordinate activities if dancing with someone else or in a group, etc. Complex new learning, neurobics, would be part of learning a new dance, and with having to coordinate movements with a partner and/or a group.

MusicMendsMinds.org is a rich source of videos, virtual events, and information on the many ways that their intergenerational efforts use music as medicine. They have had activities on four continents. PlayingForChange.org is also a rich source of videos of musicians in *many* countries bringing these Music and Dance benefits to the World in the interest of promoting peace. Youth in their videos exuberantly express similar concepts as have been documented in research.

Rudy Tanzi, PhD, is a Harvard neuroscientist in whose lab the first Alzheimer's gene was identified. His TEDx talk[48] on that topic is informative and ends with the very compelling *Alzheimer's Song* which he wrote and performs with Chris Mann singing this very heart-touching song. Dr. Tanzi is also a gifted musician and composer with his music available[49] at *RUDY TANZI – NIM* where he posts songs which he composed and performs on piano. These can be played as background music in any setting that could appreciate the positive ways that simply hearing music enhances neuroplasticity. His work with MusicMendsMinds.org is much appreciated; we all agree that music is medicine.

Another wise perspective is provided by this Harvard researcher:

> "As a scientist, the most important thing you can do is learn from your mistakes."

> **Rudolph E. Tanzi, Ph.D.**
> *Professor at Harvard Medical School*
> *Director, Genetics and Aging Research Unit,*
> *Massachusetts General Hospital*

Neurobics, complex new learning, is a major component of Music and Dance. Neurobics also can take many forms. The ACTIVE Study by Karlene Ball and associates[21,22] continued to document benefits of reasoning training and computerized speed of processing training even ten years after the beginning of that research.[50]

Gratitude comes easily. It is one of many positive emotions that can add to our personal brain health; it is also so easily shared that it is high on the list of ways we can contribute to the brain health of all people we are associated with and that we care for. We are gifted to be alive in the age of technology. More than ever we now have the capacity to use our human reasoning, judgment, imagination, creativity, and problem solving to influence the use of artificial intelligence for analysis of variance, pattern recognition, image analysis, automation, and information processing[5] of real-time neuroscience data to hack our neuroplasticity for our best possible brain chemistry, architecture, and performance. It bears repeating that artificial intelligence can expedite how we can bring these evidence-based interventions for increasing neuroplastic benefits to more global citizens.

> "Perpetual optimism is a force multiplier."

> **Colin Powell**
> *United States Army officer*
> *65th United States Secretary of State 2001 to 2005*

The Positive Psychology perspective of this writing is used because (1) positive emotions increase the neurochemicals in our brains that promote healthy brain functioning and healthy brain aging; (2) emotions are contagious and our sharing of our positive emotions might add to the positive impact on the neuroplasticity of those people that we associate with and those we care for; (3) evolving neuroscience continues to refine and expand on how we can be instrumental in "Enriching Heredity"[3] AND (4) we are gifted to be alive in the age of technology when we can exert positive influences on the

Flynn[1,2] effect with the urgent goal of making our World a better place for ALL through our Intelligent Neuroplasticity Revolution.

The finding of significant increases in human intelligence "from one generation to another" during the 20th century, which Flynn has documented in "at least 34 countries," suggests that humans have been doing several beneficial things for neuroplasticity.[1,2] This Flynn effect also invites us to harness and hasten these increases in human intelligence by using artificial intelligence in being responsibly and creatively proactive in harvesting the spectrum of evidence-based interventions, culturally tailoring them, and bringing their potential benefits to our global citizens.

> True intuitive expertise is learned from prolonged experience, with good feedback on mistakes.
>
> *Daniel Kahneman*

Artificial Intelligence from Prenatal to Supercentenarians

First, let us appreciate the research of Michael Ramscar and colleagues[51] who wrote "The Myth of Cognitive Decline." At the very least, their research shows that the nuances of changes in memory are related to the accumulation of continually increasing amounts of knowledge.

When I taught that concept, the example I gave was formerly being able to carry every important paper in my briefcase; however, somehow that exploded into 27 large metal file drawers full of important papers. Finding anything requires knowing which filing cabinet to go to, which drawer to open, and which file folder to pull out. Ramscar's research[51] provides a more succinct and elegant explanation of cognitive changes across time. He found that the continual accumulation of increasing amounts of information in your brain forces different search options. His voice of Positive Psychology can empower us to celebrate senior successes.

Next, let us raise a cup of cocoa (sweetened with stevia) and a chorus of celebration that we were born in the age of technology. The artificial intelligence revolution can give us, with a click and in a flash: (1) the decades of research on the remarkable quality of life of the Okinawan centenarians; (2) research from the time and labs of Marian Diamond,[3] the mother of neuroplasticity, identifying factors for *Enriching Heredity*, (3) Karlene Ball's application[21,22] of computer-based cognitive training to enhance brain functions, independence, and quality of life, which includes some improvements that endure even as long as ten years and may be as much as 2.5 standard deviations better than average; (4) the work of Karlene Ball[21,22] and Michael Merzenich[23] leading them to say that neurocognitive decline can be prevented, delayed, and/or reversed; (5) studies of Fred Gage that brought us proof of neurogenesis in humans as well as examples of (a) how to increase neurogenesis and (b) how to increase by a multiple of five the survival of our new brain cells; (6) findings by Richard Davidson[52] on how to change brain architecture and performance by thought alone; AND (7) advances in noninvasive ways to appreciate the measurement of changes in human brain chemistry, architecture, and performance in ways that have begun to flesh out the facets of Marian Diamond's earlier predictions that humans can enjoy multiple ways of "Enriching Heredity" *at any age.*

And that's just the beginning of a list that could go on for who knows how long!!

At the same time, these advances increasingly highlight that the complexity of the human brain will afford us eons more for conducting rewarding research. This is perhaps captured most elegantly by the statement in a Fred Gage paper: "In short, while the technological advances of the last couple of decades have revealed most of what we know about adult neurogenesis, we may yet learn that we still know very little."

Thus, even while our most well thought out and designed strategies for today empower us, we will constantly want to remember that it is a gift to have artificial intelligence to assist in searches of evolving neuroscience that bring us evidence that might seriously advance our protocol. Artificial intelligence with human creativity and guidance can help us hack neuroplasticity by constantly revisiting, updating, and refining the multiple ways we will influence *"Enriching Heredity"* for our unique version of Ideal Aging as we also celebrate our ready access to evolving neuroscience. Artificial intelligence with human intelligence can empower efforts from prenatal to and through supercentenarians for ALL.

ALL LANDS says it all. **ALL** Brain Power *AT ANY AGE* could be enhanced by our Intelligent Neuroplasticity Revolution.

L – LOVE for the HEALTH of It
A – Aerobics, Balance, Strength, AND Flexibility
N – Neurobics
D – Diet to Reduce Inflammation and Nourish Your Neuroplasticity
S – Sleep of Sufficient Quality for Essential Body Work to Clean and Feed

This template to assist in remembering neuroscience on "Enriching Heredity" is at once simplistic and at the same time complex. It is simple to facilitate establishing your unique protocol. It is complex in the many ways these categories interact.

ALL our efforts begin with Love, in part because love of self will increase your likelihood of affording yourself the very best of care; also, because love and appreciation of others is so freely and easily shared. As I walk the hallways of Harborview Medical Center, there are many people that can be thanked in passing by. "Thank you for bringing me my lunch" said to the individual pushing the heavy cart of lunch trays for our patients elicits chuckles and laughter which is good for both of our brains. Love and Using Adversity is constantly essential and can have immense positive neuroplastic purchase.

The second reason for LOVE as our beginning is that positive emotions are free and easy to share; in doing so, we can enhance the brain health and general health and well-being of others. Positive Psychology has given us much research to show the benefits of these positive emotions which are contagious. That means that your own positive emotions could benefit those around you as well as yourself.

Some say, it is not possible to give without receiving in return; in the case of positive emotions, that is a good thing since our positive emotions can contribute to physical, mental, emotional, cognitive, career, business, and brain health for ALL.

> Do unto others 20% better than you would expect them to do unto you, to correct for subjective error.

> **Linus Pauling, PhD in Chemistry**
>
> *California Institute of Technology*
>
> *Nobel Prize for chemistry in 1954*
>
> *Nobel Peace Prize in 1963*

Another reason this discussion and a chapter of this book begin with LOVE is that you are the only person in your universe that will spend 24/7 with yourself. Thoughts are a choice. When we appreciate the many ways that our positive emotions can enhance our brain chemistry, architecture, and performance, we can be more highly motivated to frame things in the positive for the benefit of our brain health and the brain health of all around us. During this pandemic and civic unrest, each of us has more than the usual amount of adversity of some form and magnitude. So much so that it can become necessary to reframe it by saying: "This is another opportunity to expand my skills" in whatever way is essential at the time.

We must remember that the adversity of some of our people is beyond imagination. For some, life is horrific. It is graphic in the quotes by Worthington and Henley in earlier pages. It surely also includes what some people experience in the unending trauma of the several severe illnesses, conditions, and their treatment, if there even is a treatment available and possible. Everyone needs several people with whom they can share their agony, maybe cry for a few minutes, and then return to being happier.

The worse the adversity the greater the need for some form of positive reframe. At the same time, remember that even brief negative emotions leave chemical footprints. That is why the second chapter is Aerobics, Balance, Strength, AND Flexibility. All of these are important for our physical health and well-being. Best benefits are realized when exercise programs factor in level of fitness. They can protect our brain by reducing the chance of falling. Aerobic exercises work off the chemistry that was generated for any necessary race across the savannah to avoid being someone's lunch; aerobics also can be more effective than antidepressants commonly prescribed for depression.

Maintaining an aerobic pace for up to an hour followed by ten minutes of weightlifting increases my energy so much in the morning that you do not want to try to come between me and getting that workout. Gratitude includes living in a condominium that has an exercise room. It was no problem that only one person was permitted in there at a time during COVID-19 because my aerobics are finished before most people in my time zone are awake.

In fairness, I must say that knowledge about the health benefits of aerobics was never enough to get me started. Nor would that help me continue. This ties into Love and Using Adversity; only with a positive reinforcement program could I even begin. It is because I label and celebrate the increased level of better energetic, emotional, cognitive, and physical performance which result from being aerobic that it has become very true that you do NOT want to try to get between me and my aerobics and weightlifting!! Since a recent study found even longer lasting benefits of increasing neurogenesis from resistance training, that has been a joy to add to my exercise routine. Aerobics will continue to be my preference because I can read and write while at an aerobic pace on a machine. So, another one of my reinforcements is reading neuroscience. Let's influence the use of artificial intelligence to develop a device that will monitor weightlifting while reading. Sounds invaluable.

Above I shared my version of why that is wise use of my time. We know that aerobic exercise can increase neurogenesis, the birth of new brain cells, in humans. Research with animals shows that only about half of the new neurons survive.

What a waste when you consider animal research showing that the best way to increase the survival rate of new brain cells as much as fivefold is complex new learning!! Human research combining aerobics and complex new learning has shown much promise.

Scientists tell us that it takes about two weeks after birth before the surviving brain cells begin to integrate. While they are young and excitable, they require an invitation to join your elegant brain. That suggests to me that the complex new learning that I love getting while on a cardio machine today is helping to integrate more of the cells that were birthed weeks ago while, simultaneously, being aerobic is increasing the birth rate of new neurons today.

I am speculating. But this speculation follows logically from what I know of science. Also, it adds to my happiness as I continue my habit of getting my aerobic exercise in the morning before most people in my time zone have gotten out of bed.

The critical point always is that you find your own enticements, entertainment, and reinforcements for doing the amount of aerobic exercise you and your healthcare provider decide is best for you. Healthcare provider guidance is essential because of both extremes: (1) damage that can accrue from too little exercise captures the attention of the media so widely that it does not need to be repeated here; AND (2) less well known is that being an endurance athlete is a risk factor for atrial fibrillation (AFib). Thus, confer with your healthcare provider. Then, incentivize.

That brings us back to the gift of **Artificial Intelligence from Prenatal to Supercentenarians.** There are multiple ways that we can use artificial intelligence to hack neuroplasticity for healthy aging across our entire lifespan.

Identify

There are many evidence-based interventions in each of the categories: Love, Aerobics, Neurobics, Diet, and Sleep. Database searches for the category of choice can be much more efficient with artificial intelligence which can also provide the articles describing these interventions for people who want clarity on such things as the scientific method, population studied, and other specifics behind the results of the study.

Personalize

Once artificial intelligence helps identify the evidence-based intervention(s) of choice, it can also tailor its use to the characteristics of the individual such as age, goals, mobility, understanding of the techniques of using the chosen intervention, and health status. It can help set strategies for achieving goals.

Prioritize

Which goals are of most importance to the individual? Are health considerations pertinent to setting priorities? Artificial intelligence can help an individual sort the personal and scientific data to set these priorities.

Incentivize

The importance of incentives cannot be emphasized enough. Knowledge alone is rarely sufficient to influence people to make the healthiest lifestyle choices. If it were, we would not be facing the health crises prominent currently.

Artificial intelligence can help identify and lengthen the list of incentives most likely to increase an individual's healthy lifestyle choices. Devices can provide positive reinforcement of beginning, completing, and repeating scheduled activities. When an individual is a predetermined amount of time late for beginning a scheduled individual, this device can provide a prompt.

Artificial intelligence can add a variety of forms of incentives. The incentives can be programmed to be unpredictable in timing and frequency.

Advise

The individual can be given data on their progress and any suggested changes for additional positive influence on hacking their neuroplasticity. If the individual chooses to share the metrics of the effectiveness of their interventions, this can be shared with their healthcare provider.

Revise

Since this data can be available in real time on the device, the individual can again use artificial intelligence to revise their goals and strategies. That can reinforce progress. It can take advantage of evolving neuroscience.

With human intelligence influencing and directing the use of artificial intelligence from prenatal to supercentenarians, we can maximize hacking neuroplasticity for the best possible healthy aging of our brains. We can be strategic in efforts to influence our healthspan to match our lifespan.

ALL LANDS can give you the template to remember important categories you will factor in when conferring with your healthcare provider about your strategies to increase your brain power *AT ANY AGE*. It is a template that affords flexibility as evolving neuroscience adds to our database on wise lifestyle choices.

After hearing Marian Diamond, the mother of neuroplasticity, speak to a Washington State Psychological Association meeting, my clinical approach shifted. When I taught the value of these interventions for physical health, the response was usually something like: "Yeh, I know. I gotta get on it." After hearing Dr. Diamond's research on driving neuroplasticity in a positive direction, I began telling clients, in the interest of full disclosure, that this evidence-based perspective in those early days primarily included research on rats; asserting that sometimes there could be a certain wisdom in acting like a rat since rats voluntarily run many meters per day; asking that they celebrate that their brain is plastic; explaining that, by the time they did everything that had been shown to improve brain plasticity, probably the only side effects would be improved general health and well-being; and further empowering them with Positive Psychology tools and perspectives. Since Dr. Diamond proved that the rats with the most toys did win, my emphasis changed to the impact of these lifestyle choices on our *brains.*

"Give me the book! I'm doing this!" was the most frequent response to this approach using Positive Psychology to focus on how we might drive brain plasticity in a positive direction for "Enriching Heredity." There was a significant increase in motivation, compliance, progress, and enjoyment of the therapy process and progress. Let this information be the wind beneath your wings of change.

I believe we have never had as much leverage for improving our brain health and our brain power as we have today. We know more about how we can define our own Ideal Aging and influence our own vigorous longevity while "Enriching Heredity," to use the title of Diamond's book, and the essence of the quotes above of several researchers have shown that she was prescient in saying humans could enjoy the same brain benefits in vigorous longevity.

In designing "Breakthrough Devices" to expedit Flexible Savings Accounts (FSA) approval and medical insurance coverage, it is essential to generate good evidence of efficacy as well as the costs benefits ratio.[53] This is particularly pertinent in assuring access for patients whose medical coverage is Medicaid and Medicare.

We can hack neuroplasticity with good judgment and problem solving in the use of artificial intelligence. Since these advances can be accomplished in ways that can simultaneously promote prosocial behaviors, our efforts have the potential to also influence peaceful coexistence. That gives us reason to hope for further progress which can be refined and accelerated with our Intelligent Neuroplasticity Revolution.

Diamond[3] wrote the book *Enriching Heredity* in which she gave great details for the scientists among us of her research showing impressive changes in brain chemistry, architecture, and performance in rats with an enriched environment in the company of other rats. Any rat that was alone in an adjoining cage, whose only brain stimulation was observation of the activities of the animals in the enriched environment, did not produce these changes in their neuroplasticity. For her vast body of research, Diamond can be seen as the mother of neuroplasticity. Diamond's book, *Magic Trees of the Mind*, addresses how applying effective interventions from her research could provide neuroplastic benefits in youth.[4]

The *very* significant neuroplastic purchases described above of the solitary addition of "TLC" to Marian Diamond's research protocol strengthens the resolve to proceed with Positive Psychology, love, and the healing power of touch for interventions going forward.

> Do unto others 20% better than you would expect them to do unto you,
>   to correct for subjective error.
>
> **Linus Pauling, PhD in Chemistry, California**
> **Institute of Technology**
> *Nobel Prize for chemistry in 1954*
> *Nobel Peace Prize in 1963*

Artificial intelligence can help us include these evidence-based interventions in our culturally tailored strategies for our Intelligent Neuroplasticity Revolution with the goals of (1) increasing the Flynn[1,2] effect; (2) celebrating the finding described in "The Myth of Cognitive Decline"[51]; and (3) enjoying better physical, mental, cognitive, emotional, career, academic, and social health. These efforts could assist international projects to prevent, delay onset, and/or reverse dementia. With strategic joint efforts, these benefits could be appreciated in more than the 34 countries documented by Flynn.[1,2]

Current neuroscience provides us with sufficient data to influence these increases in brain health and brain power beginning at the first moments of life and for the duration of a vigorous longevity.

Celebrating serious centenarian successes can serve to increase the wind beneath our wings of change. Robert Marchand[13] set world records for track bicycling at the ages of 100, 103, and 105 while also giving the first proof of the capacity to improve cardiovascular functioning in a centenarian. There are several other examples of stellar centenarian successes given herein. Our decrease in incidence and prevalence of dementia as well as the finding of increasing human intelligence in 34 countries can also fuel the flight of our evolving personal best as well as our motivation to facilitate such progress for ALL. One of our best neurobics is music and this value increases with the addition of dance. Ongoing research continually provides evidence that combining the multiple evidence-based interventions can increase our benefits.

MusicMendsMinds.org is an example of bringing the benefits of music as medicine to our global citizens. PlayingForChange.org similarly makes our World a better place with music and dance in the interest of promoting peaceful coexistence.

> *Happiness is when what you think, what you say, and what you do are in harmony.*

> **Mahatma Gandhi**

I taught hope through neuroplasticity, your ability to change your brain by learning new behavioral default options (habits), to learn new behaviors that could serve you better in your present situation, for example, being aerobic rather than sedentary. I taught that that could be a good thing if that would also modify the lollipops on your dendrites handling your new behaviors as it did in rats. Lollipops are one kind of spine that stores memories as they build on the dendrite of brain cell. In any case, it is a nice visual to focus on while you work through sufficient rehearsal of the new behaviors until they become your new and healthy habits, i. e., while you are establishing new default options.

Most people appreciated that as a model of hope and personal empowerment. So much so that I would get handmade greeting cards such as this one in Figure 9.2 below.

Our new mantra: Use them and they will grow.

Figure 9.2
This is a photo of a Christmas card I received from a couple who were grateful for all I had taught them about neuroplasticity, including that memories are stored on spines (some of which are called lollipop spines) on the dendrites of your brain cells.

A Christmas Wish For You, Joyce:

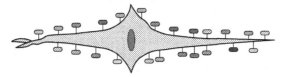

May You Have Copious Lollipops On Your Dendrites

May you have much joy in doing so. What is ahead might be continually improving your brain health and brain power through increasing brain volume in strategic and chosen areas while enjoying complex new learning in the specialty of your choice with the goal of fine tuning your brain architecture even on the level of the molecules of communication such that the trend of upwardly evolving human intelligence globally establishes new heights on the charts for a human communion infused with happiness in peaceful coexistence.

You can influence that trend. We can work *and* play at this together.

In those rare moments that you feel overdosed with new science in this information explosion, pause to celebrate how much you have already learned and changed. Since you have read this far, it is possible that your personal best might be floating up with gains in many realms: social; personal; commercial; financial; intellectual; memory; speed; philosophical perspective; happiness … any others?

I wish you the very best in your Intelligent Neuroplasticity Revolution of using artificial intelligence in hacking neuroplasticity to help enhance your healthy aging brain! Enjoy using artificial intelligence to obtain and sort through increasingly empowering research, celebrating your progress, and being a role model for using evidence-based interventions in your personalized path of *Enriching Heredity.*

Much of life takes a team. Working together, we can make our World a better place for ALL.

References

1. Flynn JR (2012). *Are We Getting Smarter? Rising IQ in the Twenty-First Century.* New York: Cambridge University Press.
2. Flynn JR. (2018). Reflections about Intelligence Over 40 Years. *Intelligence,* 70: 73–83. https://doi.org/10.1016/j.intell.2018.06.007
3. Diamond MC. (1988). *Enriching Heredity: The Impact of the Environment on the Anatomy of the Brain.* New York: The Free Press.
4. Diamond MC & Hopson J. (1999). *Magic Trees of the Mind. How to Nurture Your Child's Intelligence, Creativity, and Health Emotions from Birth through Adolescence.* New York: Plume.
5. Lawry T. (2023). *Hacking Health Care: How AI and the Intelligence Revolution Will Reboot an Ailing System.* New York: Routledge.
6. de Godoy LL, Alves CAPF, Saavedra JSM, Studart-Neto A, Nitrini R, Leite CdC & Bisdas S. (2021). Understanding Brain Resilience in Superagers: A Systematic Review. *Neuroradiology,* 63: 663–683. https://doi.org/10.1007/s00234-020-02562-1
7. de Godoy LL, Studart-Neto A, Wylezinska-Arridge M, Tsunemi MH, Moraes NC, Yassuda MS, Coutinho AM, et al. (2021). The Brain Metabolic Signature in Superagers Using In Vivo1H-MRS: A Pilot Study. *AJNR Am J Neuroradiol,* 42(10): 1790–1797. https://doi.org/10.3174/ajnr.A7262
8. de Godoy LL, Studart-Neto A, de Paula DR, Green N, Halder A, Arantes P, Chaim KT, et al. (2023). Phenotyping Superagers Using Resting-State fMRI. *Am J Neuroradiol.* https://www.ajnr.org/content/early/2023/03/16/ajnr.A7820
9. Mohammadiarvejeh P, Klinedinst BS, Wang Q, Li T, Larsen B, Pollpeter A, Moody SN, Willette SA, Mochel JP, Allenspach K, Hu G & Willett AA. (2023). Bioenergetic and Vascular Predictors of Potential Super-Ager and Cognitive Decline Trajectories—A UK Biobank Random Forest Classification Study. *GeroScience,* 45: 491–505. https://doi.org/10.1007/s11357-022-00657-6
10. Holstege H, Beker N, Dijkstra T, Pieterse K, Wemmenhove E, Schouten K, et al. (2018). The 100-plus Study of Cognitively Healthy Centenarians: Rationale, Design and Cohort Description. *Eur J Epidemiol,* 33: 1229–1249. https://doi.org/10.1007/s10654-018-0451-3
11. Mizutani T & Shimada H. (1992). Neuropathological Background of Twenty-Seven Centenarian Brains. *J Neurol Sci,* 108: 168–177. https://doi.org/10.1016/0022-510X(92)90047-O
12. Jiang J, Liu T, Crawford JD, Kochan NA, Brodaty H, Sachdev PS, et al. (2020). Stronger Bilateral Functional Connectivity of the Frontoparietal Control Network in Near-Centenarians and Centenarians without Dementia. *Neuroimage,* 215: 116855. https://doi.org/10.1016/j.neuroimage.2020.116855
13. Billat V, Dhonneur G, Mille-Hamard L, Moyec LL, Momken I, Launay T, et al. (2017). Case Studies in Physiology: Maximal Oxygen Consumption and Performance in a Centenarian Cyclist. *J Appl Physiol,* 122: 430–434. https://doi.org/10.1152/japplphysiol.00569.2016
14. Andersen-Ranberg K, Vasegaard L & Jeune B (2001). Dementia Is Not Inevitable: A Population-Based Study of Danish Centenarians. *J Gerontol B: Psychol Sci Soc Sci,* 56: 152–159. https://doi.org/10.1093/geronb/56.3.P152

15. Perls TT. (2004). Centenarians Who Avoid Dementia. *Trends Neurosci*, 27: 633–636. https://doi.org/10.1016/j.tins.2004.07.012
16. Jopp DS, Park M-KS, Lehrfeld J & Paggi ME. (2016). Physical, Cognitive, Social and Mental Health in Near-Centenarians and Centenarians Living in New York City: Findings from the Fordham Centenarian Study. *BMC Geriatr*, 16: 1. https://doi.org/10.1186/s12877-015-0167-0
17. Qiu C & Fratiglioni L. (2018). Aging Without Dementia Is Achievable: Current Evidence from Epidemiological Research. *J Alzheimer Dis*, 62: 933–942. https://doi.org/10.3233/JAD-171037
18. Motta M, Ferlito L, Magnolf SU, Petruzzi E, Pinzani P, Malentacchi F, et al. (2008). Cognitive and Functional Status in the Extreme Longevity. *Arch Gerontol Geriatr*, 46: 245–252. https://doi.org/10.1016/j.archger.2007.04.004
19. Silver MH, Jilinskaia E & Perls TT. (2001). Cognitive Functional Status of Age-Confirmed Centenarians in a Population-Based Study. *J Gerontol Psychol Sci*, 56B: P134–P140 DOI: 10.1093/geronb/56.3.p134
20. Satizabal CL, Beiser AS, Chouraki V, Chene G, Dufouil C & Seshadri S. (2016). Incidence of Dementia Over Three Decades in the Framingham Heart Study. *N Engl J Med*, 374: 523–532. https://doi.org/10.1056/NEJMoa1504327
21. Ball KK, Berch DB, Helmers KF, Jobe JB, Leveck MD, Mariske M, et al. (2002). Effects of Cognitive Training Interventions with Older Adults: A Randomized Controlled Trial. *JAMA*, 288: 2271–2281. https://doi.org/10.1001/jama.288.18.2271
22. Ball KK, Ross LA, Roth DL & Edwards JD. (2013). Speed of Processing Training in the Active Study: Who Benefits? *J Aging Health*, 25: 65S–84S. https://doi.org/10.1177/0898264312470167
23. Mahncke H, Bronstone A & Merzenich MM. (2006). Brain Plasticity and Functional Losses in the Aged: Scientific Bases for a Novel Intervention. *Prog Brain Res*, 157: 81–109. https://doi.org/10.1016/S0079–6123(06)57006-2
24. Diamond MC. (2001). Response to the Brain of Enrichment. *An Acad Bras Cienc*, 73: 211–220. https://doi.org/10.1016/B0-08-043076-7/03626-3
25. Angevaren M, Aufdemkampe G, Verhaar HJJ, Aleman A & Vanhees L. (2008). Physical Activity and Enhanced Fitness to Improve Cognitive Function in Older People Without Known Cognitive Impairment. *Cochrane Database Syst Rev*, 2: CD005381. https://doi.org/10.1002/14651858.CD005381.pub3
26. Baker LD, Frank LL, Foster-Schubert K, Green PS, Wilkinson CW, McTiernan A, et al. (2010). Effects of Aerobic Exercise on Mild Cognitive Impairment: A Controlled Trial. *Arch Neurol*, 67: 71–79. https://doi.org/10.1001/archneurol.2009.307
27. Burzynska AZ, Jiao Y, Knecht AM, Fanning J, Awick EA, Chen T, et al. (2017). White Matter Integrity Declined Over 6-Months, But Dance Intervention Improved Integrity of the Fornix of Older Adults. *Front Aging Neurosci*, 9: 59. https://doi.org/10.3389/fnagi.2017.00059
28. Edwards JD, Fausto BA, Tetlow AM, Crorna RT & Valdés EG. (2018). Systematic Review and Meta-analyses of Useful Field of View Cognitive Training. *Neurosci Biobehav Rev*, 84: 72–91. https://doi.org/10.1016/j.neubiorev.2017.11.004
29. Erickson KI, Leckie RL & Weinstein AM. (2014). Physical Activity, Fitness, and Gray Matter Volume. *Neurobiol Aging*, 35: 20–28. https://doi.org/10.1016/j.neurobiolaging.2014.03.034

30. Erickson KI, Voss MW, Prakash RS, Basak C, Szabo A, Chaddock L, et al. (2011). Exercise Training Increases Size of Hippocampus and Improves Memory. *Proc Natl Acad Sci U S A*, 108: 3017–3022. https://doi.org/10.1073/pnas.1015950108

31. Jessberger S & Gage FH. (2014). Adult Neurogenesis: Bridging the Gap between Mice and Humans. *Trends Cell Biol*, 24: 558–563. https://doi.org/10.1016/j.tcb.2014.07.003

32. Lojovich JM. (2010). The Relationship Between Aerobic Exercise and Cognition: Is Movement Medicinal? *J Head Trauma Rehabil*, 25: 184–192. https://doi.org/10.1097/HTR.0b013e3181dc78cd

33. Larson EB. (2008). Physical Activity for Older Adults at Risk for Alzheimer Disease. *JAMA*, 300: 1077–1079. https://doi.org/10.1001/jama.300.9.1077

34. Nagamatsu LS, Flicker L, Kramer AF, Voss MW, Erickson KI, Hsu CL, et al. (2014). Exercise Is Medicine, for the Body and the Brain. *Br J Sports Med*, 48: 943–944. https://doi.org/10.1136/bjsports-2013–093224

35. Niemann C, Godde B & Voelcker-Rehage C. (2016). Senior Dance Experience, Cognitive Performance, and Brain Volume in Older Women. *Neural Plast*: 9837321. https://doi.org/10.1155/2016/9837321

36. Pereira AC, Huddleston DE, Brickman AM, Sosunov AA, Hen R, McKhann GM, et al. (2007). An in Vivo Correlate of Exercise-Induced Neurogenesis in the Adult Dentate Gyrus. *Proc Natl Acad Sci U S A*, 104: 5638–5643. https://doi.org/10.1073/pnas.0611721104

37. Ryan SM & Nolan Y. (2016). Neuroinflammation Negatively Affects Adult Hippocampal Neurogenesis and Cognition: Can Exercise Compensate? *Neurosci Biobehav Rev*, 61: 121–131. https://doi.org/10.1016/j.neubiorev.2015.12.004

38. Shaffer J. (2016). Neuroplasticity and Clinical Practice: Building Brain Power for Health. *Front Psychol*, 7: 1118. https://doi.org/10.3389/fpsyg.2016.01118

39. Ritchie K & Kildea D. (1995). Is Senile Dementia "Age-Related" or "Ageing-Related"? – Evidence from Meta-analysis of Dementia Prevalence in the Oldest Old. *Lancet*, 346: 931–934. https://doi.org/10.1016/S0140-6736(95)91556-7

40. Beker N, Sikkes SAM, Hulsman M, Schmand B, Scheltens P & Holstege H. (2019). Neuropsychological Test Performance of Cognitively Healthy Centenarians: Normative Data from the Dutch 100-plus Study. *J Am Geriatr Soc*, 67: 759–767. https://doi.org/10.1111/jgs.15729

41. Beker N, Sikkes SAM, Hulsman M, Tesi N, van der Lee SJ, Scheltens P, et al. (2020). Longitudinal Maintenance of Cognitive Health in Centenarians in the 100-plus Study. *JAMA Netw Open*, 3: e200094. https://doi.org/10.1001/jamanetworkopen.2020.0094

42. Horowitz AM, Fan X, Bieri G, Smith LK, Sanchez-Diaz CI, Schroer AB, et al. (2020). Blood Factors Transfer Beneficial Effects of Exercise on Neurogenesis and Cognition to the Aged Brain. *Science*, 369: 167–173. https://doi.org/10.1126/science.aaw2622

43. Baker LD, Bayer-Carter JL, Skinner J, Montine TJ, Cholerton BA, Callaghan M, Leverenz JB Walter BK, Tsai E, Postupna N, Lampe J & Craft S. (2012). High-Intensity Physical Activity Modulates Diet Effects on Cerebrospinal β-Amyloid Levels in Normal Aging and Mild Cognitive Impairment. *J Alzheimers Dis*, 28(1): 137–146. https://doi.org/10.3233/JAD-2011-111076.

44. Baker LD. (2015). Aerobic Exercise Reduces CSF Levels of Phosphorylated Tau in Older Adults with MCI. Alzheimer's Association International Conference 2015 presentation.
45. Wang S, Yin H, Meng X, Shang B, Meng Q, Zheng L, et al. (2019). Effects of Chinese Square Dancing on Older Adults with Mild Cognitive Impairment. *Geriatr Nurs*, 41: 290–296. https://doi.org/10.1016/j.gerinurse.2019.10.009
46. Rehfeld K, Müller P, Aye N, Schmicker M, Dordevic M, Kaufmann J, et al. (2017). Dancing or Fitness Sport? The Effects of Two Training Programs on Hippocampal Plasticity and Balance Abilities in Healthy Seniors. *Front Hum Neurosci*, 11: 305. https://doi.org/10.3389/fnhum.2017.00305
47. Rehfeld K, Luders A, Hoekelmann A, Lessmann V, Kaufmann J, Brigadski T, et al. (2018). Dance Training Is Superior to Repetitive Physical Exercise in Inducing Brain Plasticity in the Elderly. *PLoS ONE*, 13: e0196636. https://doi.org/10.1371/journal.pone.0196636
48. Tanzi R. Curing Alzheimer's with Science and Song | Rudy Tanzi & Chris Mann | TEDxNatick (youtube.com).
49. Tanzi R. https://www.n1m.com/rudytanzi
50. Rebok GW, Ball KK, Guey LT, Jones RN, Kim H-Y, King JW, et al. (2014). Ten-905 Year Effects of the Advanced Cognitive Training for Independent and Vital Elderly Cognitive Training Trial on Cognition and Everyday Functioning in Older Adults. *J American Geria Soc*, 62: 16–24. https://doi.org/10.1111/jgs.12607.
51. Ramscar M, Hendrix P, Shaoul C, Milin P & Baayen H. (2014). The Myth of Cognitive Decline: Non-linear Dynamics of Lifelong Learning. *Top Cogn Sci*, 6: 5–42. https://doi.org/10.1111/tops.12078
52. Davidson RJ & Schuyler BS (2015). Neuroscience of Happiness in WORLD HAPPINESS REPORT 2015 available at WorldHappiness.report
53. Kadakia KT, Kramer DB & Yeh RW (2023). Coverage for Emerging Technologies – Bridging Regulatory Approval and Patient Access. *N Engl J Med*, 89: 2021–2024. https://doi.org/10.1056/NEJMp2308736

10

Our Intelligent Neuroplasticity Revolution in Business, Academia, Healthcare, Governments, and Philanthropy

My life and service to others changed significantly after hearing Marian Diamond,[1] the Mother of Neuroplasticity, describe her research. Complex new learning for your neurons as seen above in Figure 10.1 is essential. Diamond said "and then we added LOVE" as Figure 10.2 above illustrates. Adding LOVE resulted in *huge brain benefits* which continued to build bigger better brains across their 50% increased lifespan. Much of what she taught I had learned and applied from a medical perspective.

Her clarity on driving neuroplasticity in a positive direction in ways that afforded her animals a 50% increase in lifespan to the equivalent of 90 human years, with "Enriching Heredity" continuing throughout their vigorous healthspan and extended lifespan, was as life-changing for me as it has been for others I have served. Diamond[1] opined that "Enriching Heredity" is possible for a human through 100 years of life.

Increasing the use of these evidence-based lifestyle choices could lead to significant financial gains and other benefits in multiple ways for more people. The options described herein for improving brain chemistry, architecture, and performance are also associated with improved general health and well-being, increasing productivity, and more prosocial behaviors which could include more prosociality and peaceful coexistence.

That could result in a healthier work environment, living situation, and reduced healthcare costs for everyone. Other significant gains could be

DOI: 10.4324/9781003462354-10

Figure 10.1
These illustrations by David Yu, MD, capture the importance of complex new learning, LOVE, and stress reduction as part of what we need to enhance our neuroplasticity.

avoiding the frequent costs of training new employees. It could entice highly qualified individuals to participate in your endeavors.

Employees, directors, managers, trainees, students, clinicians, philanthropists, volunteers in service to others ... these humans are our greatest resources, our most significant assets. That highlights the significance of the role of the decision-makers and human resources department in all these settings.

Decision-makers and human resources directors who advocate for the use of the evidence-based lifestyle choices described herein have the potential to (1) influence increased productivity; (2) notice enhanced brain power;

Figure 10.2
These illustrations by David Yu, MD, capture the importance of complex new learning, LOVE, and stress reduction as part of what we need to enhance our neuroplasticity.

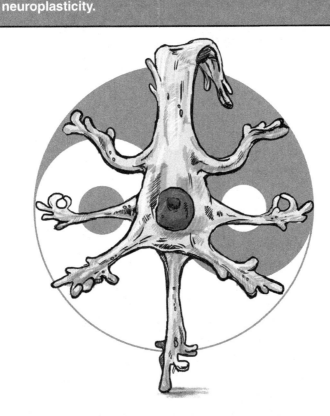

(3) observe improved physical health and well-being; (4) influence reduced healthcare costs; (5) see more prosocial behaviors; and (6) appreciate more civility in human interactions.

The risks in *not* advocating for the use of evidence-based lifestyle choices written herein include the antithetical of all the above for the short term as well as across the lifespan. That's why we need to enjoy the Intelligent Neuroplasticity Revolution on the job and off.

My reasons for gratitude include the amount of time I spend on my feet on my job as a court evaluator at Harborview Medical Center. I climb many flights of stairs to get to my office; walk many city blocks to and from the several hospital units; then go down several flights of stairs backward (hanging on tightly to the railing) to improve my balance. If your job requires sitting a lot, what options are available at break time; before work; and off the job?

Combining activities can have broader purchase. Autophagy as described herein is influenced by the ratio of energy consumed and energy used; it can reduce aging of brain cells and be neuroprotective.[2] It's conceivable that eating a healthy diet that includes reducing calories while also increasing physical exercise might increase autophagy.

Failing to increase healthy lifestyle choices as influenced by this evolving neuroscience on enhancing neuroplasticity leaves our population at risk for early onset of dementia.[3] It could also result in less productivity and less prosocial behaviors from infancy across the lifespan.

As the Fortune Magazine article[4] describes about the "5 huge governance missteps" in the OpenAI firing of Sam Altman, corporate boards need to "protect or represent shareholders" in overseeing the mission of the entity; boards need to have sufficient members with pertinent experiences; the board chair needs to be involved in decisions; major announcements need to be made in a time and space that affords employees ample opportunity for community discussions; investors need to be included in discussions for building consensus; and putting the mission of the entity at risk could undermine the goals and future of the entity.

That makes the forms of positive communication discussed in an earlier chapter of this book very pertinent in corporate and other settings. For example, remaining open to listen to the perspectives of all involved parties empowers everyone. Forgiving differences; reframing perspectives associated with anxiety and/or depression; expressing gratitude often to many people; inducing appropriate laughter that *celebrates* individuals (never put anyone down); promoting prosociality in the interest of global well-being; supporting volunteer service to others; providing social support; all of these are important and are just the beginning of that list. Including all these positives in your corporate culture constantly could reap benefits for all concerned.

Our brains are plastic. That's the best news.

Evolving neuroscience is fluid; that is a very good thing. One of the many gifts of being alive in the age of technology is that results are available to us closer to real time. Artificial intelligence can expedite our searches of this database, can help us culturally tailor its use, and can monitor neuroplastic purchase. Since even gold standard random controlled trials are still only a probability statement about the population studied, we have several degrees of freedom in what and how we, in conferring with our healthcare provider, design goals and strategies for our own unique version of "*Enriching*

Heredity"[1] while we use artificial intelligence to continually refine our use of the evolving neuroscience database and our neurobiological responses to using these lifestyle choices described in evolving research.

In addition to what we can learn from the skills, lifestyle choices, attitudes, and wisdom of our SuperAgers, positive agers, centenarians, and supercentenarians, we are obligated to address how ALL people of ALL ages can benefit from evolving neuroscience on various ways to maximize *"Enriching Heredity."*[1] The goal in our Intelligent Neuroplasticity Revolution is hacking neuroplasticity to enhance healthy brain aging for best brain power for ALL people throughout a vigorous longevity such that the individual's brain healthspan matches their lifespan. This writing provides the neuroscience related to how to do so in ways that can simultaneously enhance physical, social, emotional, neurobiological, and prosocial healthy aging.

This is of increasing importance. As our population ages around the globe, we can increase the extent to which the oldest old contribute to, as well as benefit from, the types and methods of evidence-based interventions that artificial intelligence can expedite refinement of, can assist in making age appropriate, and can develop in wearables and nearables. These personally and culturally tailored devices can prompt for, monitor use of, incentivize, be responsive to the individual's current level of functioning, and give tailored positive reinforcement for all involvement and progress.

It is prime time for humans to apply our reasoning, judgment, imagination, creativity, and problem solving to harness and influence the use of artificial intelligence for variance analysis, pattern recognition, image analysis, automation, and information processing.[5] We can influence this use of artificial intelligence to increase use of, and brain benefits with, the multitude of evidence-based interventions that can hasten the Flynn[6,7] effect of increasing human intelligence in ways that could simultaneously promote healthy aging with prosocial behaviors. Artificial intelligence with human oversight can expedite and refine our Intelligent Neuroplasticity Revolution.

Governments, businesses, academia, communities, philanthropic leaders, and personal decision-makers need to share and use this information in ways best suited to their situation and strategic healthy aging goals. Artificial intelligence with human direction can assist in cultural tailoring and in reflecting the setting, whether business, academia, healthcare, communities, or homes, to maximize the potential benefits of our work, our quality of life, and our financial savings related to healthcare costs.

With artificial intelligence, we can tailor wearables and nearables which are personalized to the extent that prompting respects and celebrates the individual patterns. For example, ten minutes after an activity was calendared, the individual would receive positive reinforcement for having begun as planned *or* prompted to do the activity if they had not yet started it. Positive reinforcement as chosen by the individual might be words, a visual such as thumbs up, or a sound such as music, applause, or birds chirping. In adopting artificial intelligence for use in devices, clinical oversight and interventions to manage risk are essential as well as quite complex.[8] "AI is not one technology but a heterogeneous group with varying liability risks. Identifying AI tools with the greatest risk can help target risk-management interventions and clinical oversight."

The critical components of the artificial intelligence-assisted wearables or nearables for increasing use of evidence-based activities for hacking neuroplasticity for health need to include but might not be limited to: (1) personalizing ALL components of the strategic plan and goals; (2) only prompting if the person is some number of minutes late to the activity they had calendared; (3) providing positive reinforcement of every step of involvement in the process, including beginning it; (4) maximizing fun, laughter, and love of doing the activities; (5) measuring speed of processing; (6) giving bonus scores for increasing speed of processing in computerized cognitive training; (7) modifying difficulty in response to the individual's current level of functioning; (8) giving positive reinforcement and bonus points for ongoing and regular use of the chosen evidence-based activities as calendared; (9) biophysiological measurements of change over time; and (10) maximum security of all data such that the individual can benefit from seeing and being positively reinforced by their progress while this same personal data is *only* available to any specific person and/or healthcare entity that the individual designates. Clinical oversight and data protection are essential.

Carefully designed wearables and nearables as described above could increase use of these evidence-based interventions by monitoring use, measuring impact, prompting for use if/when a person is behind schedule in following their plan, and providing the form of positive reinforcement that increases the pleasure in doing the specific activity. That positive reinforcement is essential because people are more likely to do what is fun, easy, portable, culturally adapted, and free or low cost. The personalized form of positive reinforcement strategically delivered is essential because that can increase the use and the benefits of these interventions.

There is increasing need for artificial intelligence to direct research and cultural tailoring of evidence-based efforts toward projects that could maximize the number of lives in our global citizens that could be impacted by potential improvements in healthy aging with enhanced brain power and prosociality. Evolving neuroscience can be expedited by judicious use of artificial intelligence. The Flynn[6,7] effect of increasing human intelligence in 34 countries can be influenced to benefit more of our global citizens by judicious use of artificial intelligence in our Intelligent Neuroplasticity Revolution.

Hacking neuroplasticity for promoting healthy aging, best possible brain health, enhanced brain power, and prosociality *at any age* is urgent. Working wisely together with artificial intelligence we can do this.

Start with:

Love

Using the example of how Marian Diamond[1] influenced significant improvements in *Enriching Heredity* by simply holding and talking with her animals, establish parameters of how this can be included in the culture of your business, classroom, and community. Handshakes on greeting others is an example of human touch we could hope would have similar neuroplastic purchase. Maybe touching elbows could have some value despite the fabric that might reduce the physical touch. Is an arm around the shoulder, like seen in the picture of the Rescuing Hug, permissible? For some, a hug as a greeting might be acceptable. Culturally tailor how this potential neuroplastic purchase of touch and talk could work in your business, classroom, and/or facility. The health insurance might also include coverage for massage to capture some of the benefits of touch as described herein.

Establish a norm for positive interactions and focus on Positive Psychology[9] in your business, community, and social interactions. In helping each other enhance our personal best, informative feedback that is most useful includes episodic compliments on a job well done as well as sharing thoughts on what an individual might consider doing differently for better outcomes. The times that I have asked to talk privately with a supervisor about a behavior that "I want you to change," they have been receptive, appreciative, and respectful in considering and/or making that change.

Express gratitude freely and frequently as fitting to your situation. Generate laughter in ways that celebrate people that you associate with, never in a way that could be taken negatively. For example, when someone walks downstairs past me in a stairwell while I am climbing upstairs, I point out that "down is easier." That often gets a chuckle, at least a smile, maybe a few words.

Have a posterboard in the hallway where employees and students can post compliments, good news, gratitude, and such like. Invite them to vote on choosing a student and/or employee of the month with the winner posted on that board later along with some of the compliments that were written about them. Voting on the employee of the year can include a party of recognition and celebration of this individual.

Working together we can influence the perspective of Positive Psychology in the development of all forms of media. For example, in all computerized training videos that reinforce speed of processing, memory, new learning, or whatever, the completion of the training can be constructive or just entertaining. If the visual images bounce, make sure that they stay intact while progress data that reinforces participation is visible. Add more positives each time there is an increase in the skill being trained.

Aerobics

Aerobic, strengthening, balance, and stretching exercises are easily done on the job. Have exercise machines available for people who arrive early, stay late, and/or use during breaks. These machines would be especially valuable if they gave a readout of heart rate during exercising. Even brief sessions of aerobic exercise are associated with cognitive and emotional gains. Strengthening exercises have similar benefits.

Businesses could give points to employees for getting the type and amount of exercise that could increase their neurogenesis, brain health, brain power, mood, productivity, and general physical health and well-being. The multiple ways that aerobic fitness influences neurocognition begin in youth with improved executive functions such as cognitive flexibility.[10] Students and employees being aerobic and/or building muscle strength could earn points toward some form of reward.

Small exercise machines that fit beneath a desk might be useful for those whose work requires extended time in a chair while on the job. If a standing desk is used, make it sturdy enough that, while a person is reading and/ or thinking through a task, they can do pushups against the desk or squats holding onto the desk.

Compliment all who take the stairs. Have sturdy railings in stairwells for all who choose to use the stairs. Railing that is continuous from top floor to bottom adds to the safety especially between landings.

MRI scans of 10,125 people between the ages of 18 and 97 years old were analyzed with deep learning models.[11] Those individuals reporting moderate to vigorous physical exercise on more days had larger volumes of several regions of their brains. They had more volume in their grey brain matter, which is where the brain cells are; in their white brain matter which protects the branches on their brain cells; in their hippocampus which is essential for memory and new learning; and in their parietal, frontal, and occipital lobes of their brains. This suggests that their moderate to vigorous exercise had a neuroprotective effect.

This is yet another good example of the value of using artificial intelligence to read details of these MRI scans more accurately and efficiently than any human could and thus provide good information for our decision-making. It adds to the data encouraging corporations, academic settings, and communities to facilitate and reinforce moderate to vigorous exercise in our global citizens.

Increasing developments in measurement methods of neuroplasticity changes with aging can increase the value of the database on these interventions. For example, learning and memory influences that increase the volume of mushroom bodies on brain cells in honeybees, which indicate significant neuroplasticity with aging, become more visible in tomography that is micro-computed when the cells are not dry samples.[12]

Neurobics

Neurobics that are paced by the individual's current level of functioning can be part of employment, classroom studies, community activities, service to others, and personal entertainment and enrichment. This new learning needs to afford enough challenge to keep the person engaged but not so much as to become discouraging.

Computerized training is a very valuable neurobic that can reinforce progress more rapidly and accurately than any human can. Recall the research cited herein showing that speed of processing training by computer was associated with benefits even ten years later. This includes benefits from using serious computer games for cognitive training with elders.[13]

Quiet music in the background surely fits in the neurobic realm given how much research exists on the positive neuroplastic purchase of simply hearing music from in utero across the entire lifespan.[14] Rudolph Tanzi is a Harvard neuroscientist in whose lab the first Alzheimer's gene was discovered. He is also a gifted musician. Recordings of Dr. Tanzi playing his own piano compositions are available at n1m.com/rudytanzi. These musical treats are free, and listening to them would not distract from deep thinking and problem solving because it is likely that these tunes have never been heard before. Tanzi agrees with MusicMendsMinds.org that music is medicine.

Diet

Diet is of huge importance and includes ... some say almost no sugar ... this author's response to reading many studies is to advocate for *no* added sugar as described in earlier chapters.

Stevia is a plant source natural sweetener with health benefits. Many of the artificial sweeteners have bad effects. There's growing research on how to successfully use stevia in such recipes as baked goods. Finishing a paper on the health benefits of stevia is on my To Do List.

Working together, we might increase the health of all concerned by influencing the food industry to use stevia as a natural, plant-based, sweetener for the health of it as described herein. We want to avoid the accumulation of advanced glycation end products (AGEs), the protein modification caused by excess glucose molecules, which influence negative health outcomes that include, but are not limited to, risk for dementia.[15]

Dietary influence might also include emphasis on plant sources of protein for the dual purchase of better cardiometabolic effects with simultaneous reduced environmental impact.[16] Further research on these factors is essential.[17]

Healthy foods need to be the only choices in the cafeterias and vending machines in corporate, academic, and community settings. These healthy food choices also need to be culturally tailored. Hopefully vendors can be influenced to increase their sources of healthy foods, including those sweetened only with stevia.

Sleep

Sleep for the joy of it. Celebrate sleeping soundly. Enjoy knowing that some of our most important work for the health of our brains is accomplished by the glymphatic system while we sleep.[18,19]

Like what happens when huge office buildings are quiet for the night, nutrients are restocked in our brains and trash such as tau is removed by our glymphatic system which works best during sleep. Feeding and cleaning for the health of our brains is an essential part of good quantity and quality of sleep.

Businesses, academicians, clinicians, and individuals of all ages need this.[20-25] This would be especially important after periods of sleep deprivation such as when individuals work a night shift or must take an overnight flight.[26,27]

Recall that when one night of sleep deprivation in rats included gentle handling to prevent sleep, neurogenesis increased significantly initially as well as 15 and 30 days later.[28] Neurogenesis is one of the most important forms of driving neuroplasticity in a positive direction. These new brain cells are essential for new learning and memory.

There is no research investigating this outcome of gentle handling to prevent sleep increasing neurogenesis in humans. Since it is not likely that similar research will be done in humans soon, some of us will speculate that "gentle handling" of humans who must miss much needed sleep might have positive neuroplastic purchase. That speculation is tempting when considering the finding of Marian Diamond[1] that holding and talking to her animals resulted in a 50% increase in their lifespan *AND* they continued *Enriching Heredity* across their vigorous longevity to the equivalent of 90 human years!!

You decide if you want to speculate this way. Corporate entities, academic settings, and healthcare facilities might consider making massage available during those times when extended sleep deprivation is essential just in case this neuroplastic purchase can be realized in humans as well. There could also be gains in including some forms of massage in the provision of health insurance.

> Hold fast to dreams
> For if dreams die
> Life is a broken-winged bird
> That cannot fly.

Langston Hughes (1901–1967)

Do It Now for the HEALTH of It

It is noteworthy that increasing neurogenesis is one of the most important ways to drive neuroplasticity in a positive direction for increasing brain power. "Brains are not computers, rather they have a reciprocal malleability of their structures which is essential for their healthy function."[29] Neurogenesis in the hippocampus plays a key role in neurocognitive functions, including memory. These new brain cells add new synapses, new connections, and life course plasticity and function.

Looking at data from 37,533 healthy people in the UK Biobank, researchers noted that the best cognitive performance was observed in those who slept seven hours each day.[30] To assess cognitive performance, these researchers looked at executive functioning estimated from tasks of speed of processing or working memory.

It is noteworthy that, although these neurocognitive skills tend to decline with age, these findings of *best executive functioning with seven hours of sleep included the 12,006 individuals who were more than 60 years old*. People who slept six-to-eight hours daily "had significantly greater grey matter volume in 46 of 139 different brain regions including the orbitofrontal cortex, hippocampi, precentral gyrus, right frontal pole and cerebellar subfields." Thus, they opined that their "findings highlight the important relationship between the modifiable lifestyle factor of sleep duration and cognition as well as a widespread association between sleep and structural brain health."

The timing is urgent for corporations, academicians, clinicians, philanthropists, communities, and individuals to put these neuroscience findings into play for humans around the world. It is indeed sobering to learn of the estimate that around our good green globe a new person is diagnosed with dementia about every seven seconds.[31] And that does not count the people who did not have access to services or those who were not assessed for early signs of dementia when they did come in for healthcare.

Brookmeyer[31] and colleagues provide a challenge, perhaps an invitation to action. It is encouraging and empowering to read their estimate that delaying the onset of dementia by as little as one year could reduce the global burden of Alzheimer's disease by as much as 9,200,000 by 2050. Working together on this, we can have a better outcome.

Neither of these estimates provides a measure of the gains in human and business resources we could appreciate by reducing cognitive decline by a mere 1% per year. By preventing, delaying, and/or reversing the onset of dementia, what impact could that have on our healthcare issues? How much could we improve the global economic burden? Reframed in the spirit of Positive Psychology, how much can we improve quality and quantity of life? How much and how rapidly can we increase the Flynn[6,7] effect of increasing human intelligence? How many more countries can we include in appreciating these benefits of driving neuroplasticity in a positive direction? What additional resources would we grasp? Where should we begin?

Since the estimate of diagnosing dementia somewhere in the world has been one new person about every seven seconds,[31] or even more frequently, the time is urgent for change. In terms of human, academic, and business resources, what gains would even one person appreciate if they reduced mental decline by a mere 1% per year?

Reframed in Positive Psychology for the healthy purchase of positive emotions, how much benefit can even one person appreciate by making lifestyle choices that could *increase* their brain volume as much as 2% *with associated better memory?* The visual image of that research finding was provided on earlier pages for the visual learners among us AND as an enticement to influence more people to give the gift of serious senior successes to themselves and, as a role model, to those they love and serve.

Is doing all evidence-based things that could have the potential for "*Enriching Heredity*"[1] worth it to you, to your students, to your employees, and to those you love and seek to serve? Can this approach add to the Flynn[6,7] effect of increasing human intelligence that has been documented in at least 34 countries? Can it improve the global statistics on dementia? Can our collective efforts improve the global statistics on improving brain health and on enhancing brain power?

An increasing number of scientists are recommending that regular aerobic and resistance exercises are the prescription of choice for preventing, slowing, and/or reversing the so-called "normal course of aging," meaning decline. Take this seriously. Apply it toward the goals of having a *better than* "normal course of aging" for the sake of yourself, your family, your employees, your students, and your business. Artificial intelligence can help guide research and personalize strategies to include improving brain health, wellness, and performance to every extent possible for a person's uniquely healthy and productive longevity.

It is my distinct opinion that this research may mean we can achieve "genius" with less than the 10,000 hours Gladwell asserts in *Outliers*. I am speculating. But I take that position because cognitive gains through computer training can be achieved with greater accuracy than by human instruction and that these gains have endured as described above.[32-34] Combining computerized cognitive training applications and physical exercise in longitudinal studies is recommended to increase data on long-term gains in memory, fitness, independence, well-being, and quality of life.[35]

> It is not the strongest of the species that survive,
> nor the most intelligent,
> but the one most responsive to change.

Charles Darwin (1809–1882)

It is now urgent to work with experts in the Intelligent Healthy Neuroplasticity Revolution to (1) measure our baseline skills and resources; (2) set goals and establish strategic plans for how we will apply evidence-based interventions for "Enriching Heredity" for (a) ourselves, (b) for those we serve, and (c) for ALL; (3) reinforce little baby steps toward goals while keeping an eye and an ear in tune with refining and expanding on what is provided in evolving neuroscience; (4) measure our progress; (5) reinforce and refine our goals and strategic interventions; AND (6) continue onward from this new baseline toward an ever evolving improvement in personal best in ourselves and for ALL whom we love, interact with, employ, teach, serve, and care about.

References

1. Diamond MC (1988). *Enriching Heredity: The Impact of the Environment on the Anatomy of the Brain.* New York: The Free Press.
2. Marzoog BA (2024). Autophagy as an Anti-senescent in Aging Neurocytes. *Curr Mol Med*, 24(2): 182–190. https://doi.org/10.2174/15665240236662301 20102718.
3. Hendriks S, Ranson JM, Peetoom K, Lourida I, Tai XY, de Vugt M, Llewellyn DJ & Köhler S. (2023). Risk Factors for Young-Onset Dementia in the UK Biobank. JAMA Neurol. https://doi.org/10.1001/jamaneurol.2023.4929.
4. Maclelllan L (2023). OpenAI made 5 huge governance missteps—here's what boards can learn from its error. *Fortune,* Article Obtained January 19, 2024, online at: Open AI's Corporate Governance Missteps Over Sam Altman Firing | Fortune.
5. Lawry T (2023). *Hacking Health Care: How AI and the Intelligence Revolution Will Reboot an Ailing System.* New York: Routledge.
6. Flynn JR (2012). *Are We Getting Smarter? Rising IQ in the Twenty-First Century.* Cambridge: Cambridge University Press.
7. Flynn JR (2018). Reflections about Intelligence Over 40 Years. *Intelligence*, 70: 73–83. https://doi.org/10.1016/j.intell.2018.06.007.
8. Mello MM & Guha N (2024). Understanding Liability Risk from Using Health Care Artificial Intelligence Tools. *N Engl J Med*, 390: 3.
9. Peterson C (2008). What Is Positive Psychology, and What Is It Not? *Psychol Today.* https://www.psychologytoday.com/us/blog/the-good-life/200805/what-is-positive-psychology-and-what-is-it-not? (Accessed online July 2, 2023).
10. Yangüez M, Raine L, Chanal J, Bavelier D & Hillman CH (2024). Aerobic Fitness and Academic Achievement: Disentangling the Indirect Role of Executive Functions and Intelligence. *Psychol Sport Exerc*, 70: 102514. https://doi.org/10.1016/j.psychsport.2023.102514.
11. Raji CA, Meysami S, Hashemi S, Garg S, Akbari N, Ahmed G, Chodakiewitz YG, Nguyen TD, Niotis K, Merrill DA & Attariwali R (2024). Exercise-Related Physical Activity Relates to Brain Volumes in 10,125 Individuals. *J Alzheimers Dis.* https://doi.org/10.3233/JAD-230740.

12. Fu S-J & Yang E-C. (2024). Neuroplasticity in Honey Bee Brains: An Enhanced Micro-computed Tomography Protocol for Precise Mushroom Body Volume Measurement. *J Neurosci Methods*, 403: 110040. https://doi.org/10.1016/j.jneumeth.2023.110040.

13. Gutiérrez-Pérez B-M, Murciano-Hueso A & de Oliveira Cardoso A-P. (2023). Use of Serious Games with Older Adults: Systematic Literature Review. *Humanit Soc Sci Commun*, 10: 939. DOI: 10.1057/s41599-023-02432-0

14. Tanzi R online music at n1m.com/rudytanzi

15. Uceda AB, Leal-Pérez F & Adrover M. (2024). Protein Glycation: A Wolf in Sweet Sheep's Clothing behind Neurodegeneration. *Neural Regen Res*, 19(5): 975–976.

16. Landry MJ, Ward CP, Cunanan KM, Durand LR, Perelman D, Robinson JL, Hennings T, Koh L, Dant C, Zeitlin A, Ebel ER, Sonnenburg ED, Sonnenburg JL & Gardner CD. (2023). Cardiometabolic Effects of Omnivorous vs Vegan Diets in Identical Twins a Randomized Clinical Trial. *JAMA Network Open*, 6(11): e2344457. https://doi.org/10.1001/jamanetworkopen.2023.44457.

17. Gardner CD, Policastro P & Wang MC (2023). Editorial: Achieving Health Equity: Sustainability of Plant-Based Diets for Human and Planetary Health. Front Public Health, 11: 1285161. https://doi.org/10.3389/fpubh.2023.1285161.

18. Hsiao W-C, Chang H-I, Hsu S-W, Lee C-C, Huang s -H, Cheng C-H, Huang C-W & Chang C-C. (2023). Association of Cognition and Brain Reserve in Aging and Glymphatic Function Using Diffusion Tensor Image-along the Perivascular Space (DTI-ALPS). *Neuroscience*, 524: 11–20. https://doi.org/10.1016/j.neuroscience.2023.04.004.

19. Iliff JJ, Wang M, Zeppenfeld DM, Venkataraman A, Plog BA, Liao Y, Deane R & Nedergaard M. (2013). Cerebral Arterial Pulsation Drives Paravascular CSF–Interstitial Fluid Exchange in the Murine Brain. *J Neurosci*, 33(46): 18190. https://doi.org/10.1523/JNEUROSCI.1592-13.2013.

20. Keil SA, Schindler AG, Wang MX, Piantino J, Silbert LC, Elliott JE, et al. (2023). Instability in Longitudinal Sleep Duration Predicts Cognitive Impairment in Aged Participants of the Seattle Longitudinal Study. *medRxiv.* preprint https://doi.org/10.1101/2023.06.07.23291098.

21. Namsrai T, Ambikairajah A & Cherbuin N. (2023). Poorer Sleep Impairs Brain Health at Midlife. *Sci Rep*, 13: 1874. https://doi.org/10.1038/s41598-023-27913-9.

22. Soehner AM, Hayes RA, Franzen PL, Goldstein TR, Hasler BP, Buysse DJ, et al. (2023). Naturalistic Sleep Patterns are Linked to Global Structural Brain Aging in Adolescence. *J Adolescent Health*, 72: 96e104. https://doi.org/10.1016/j.jadohealth.2022.08.022.

23. Tsiknia AA, Parada H, Banks SJ & Reas ET. (2023). Sleep Quality and Sleep Duration Predict Brain Microstructure among Community-Dwelling Older Adults. *Neurobiol Aging*, 125: 90–97. https://doi.org/10.1016/j.neurobiolaging.2023.02.001.

24. Uji M & Tamaki M (2023). Sleep, Learning, and Memory in Human Research Using Noninvasive Neuroimaging Techniques. *Neurosci Res*, 189: 66–74. https://doi.org/10.1016/j.neures.2022.12.013.

25. Yiallourou SR, Cribb L, Cavuoto MG, Rowsthorn E, Nicolazzo J, Gibson M, et al. (2024). Association of the Sleep Regularity Index With Incident Dementia and Brain Volume. *Neurology*®, 102: e208029. https://doi.org/10.1212/WNL.0000000000208029.

26. Peng Z, Hou Y, Xu L, Wang H, Wu S, Song T, Shao Y & Yang Y (2023). Recovery Sleep Attenuates Impairments in Working Memory Following Total Sleep Deprivation. *Front Neurosci*, 17: 1056788. https://doi.org/10.3389/fnins.2023.1056788.

27. Yan F-X, Lin J-L, Lin J-H & Lin Y-J. (2023). Altered Dynamic Brain Activity and Its Association with Memory Decline after Night Shift-Related Sleep Deprivation in Nurses. *J Clin Nurs*, 32: 3852–3862. https://doi.org/10.1111/jocn.16515.

28. Zucconi GG, Cipriani S, Balgkouranidou I & Scattoni R. (2006). "One Night" Sleep Deprivation Stimulates Hippocampal Neurogenesis. *Brain Res Bull*, 69(4): 375–381.

29. Kempermann G, Song H & Gage FH. (2023). Adult Neurogenesis in the Hippocampus. *Hippocampus*, 33: 269–270. https://doi.org/10.1002/hipo.23525.

30. Tai XY, Chen C, Manohar S & Husain M. (2022). Impact of Sleep Duration on Executive Function and Brain Structure. *Commun Biol*, 5: 201. https://doi.org/10.1038/s42003-022-03123-3

31. Brookmeyer R, Johnson E. Ziegler-Graham K & Arrighi, M. (2007). Forecasting the Global Burden of Alzheimer's Disease. *Alzheimers Dement*, 3: 186–191.

32. Ball KK., Berch DB, Helmers KF, Jobe JB Leveck MD Mariske M, et al. (2002). Effects of Cognitive Training Interventions with Older Adults: A Randomized Controlled Trial. *JAMA*, 288: 2271–2281. https://doi.org/10.1001/jama.288.18.2271.

33. Ball KK, Ross LA, Roth DL & Edwards JD (2013). Speed of Processing Training in the Active Study: Who Benefits? *J Aging Health*, 25: 65S–84S. https://doi.org/10.1177/0898264312470167.

34. Mahncke H, Bronstone A & Merzenich MM (2006). Brain Plasticity and Functional Losses in the Aged: Scientific Bases for a Novel Intervention. *Prog Brain Res*, 157: 81–109. https://doi.org/10.1016/S0079-6123(06)57006-2.

35. Silva AF, Silva RM, Murawska-Ciałowicz E, Zurek G, Danek N, Cialowicz NM, Carvalho J & Clemente FM (2024). Cognitive Training with Older Adults Using Smartphone and Web-Based Applications: A Scoping Review. *J Prev Alz Dis*, 11(3): 693–700 . https://doi.org/10.14283/jpad.2024.17.

Index

Note: *Italic* page numbers refer to figures.

Printed in the United States
by Baker & Taylor Publisher Services